Mghaiouini Redouane
Monkade Mohamed
EL Bouaria Abdeslam

Magnetic water, impact and perspectives

Mghaiouini Redouane
Monkade Mohamed
EL Bouaria Abdeslam

Magnetic water, impact and perspectives

ScienciaScripts

Imprint

Cover image: www.ingimage.com

This book is a translation from the original published under ISBN 978-620-2-54651-5.

Publisher:
Sciencia Scripts
is a trademark of
International Book Market Service Ltd., member of OmniScriptum Publishing Group
17 Meldrum Street, Beau Bassin 71504, Mauritius
Printed at: see last page
ISBN: 978-620-3-59732-5

Signing session

To my father

To my mother

To my sisters and brother

Contents

List of Abbreviations

MW: Magnetised water

NMW : Non-magnetised water

FTIR : Fourier Transform Infrared

EMF : Electromagnetic field

TW : Tap water
MW : Magnetised water
tm : Magnetisation time
DW : Distilled water
MTW : Magnetic field treated water
TDS : Total soluble solids
FA : Fly ash
NF EN 196-1 : European standard for mortars
CPJ 45 : Portland cement with an average 28-day strength of between 35 and 55 MPa
NM 10.1.004 : Moroccan Rejection Standard for the pulp, paper and paper industry
Cardboard
NF P18-560 : AFNOR standard for grain size
EN 196-1 : European standard for mortars
BA : Fireplace ash
RX : X-rays
KOH : Potassium hydroxide
OM : Organic matter
THC : Total humic compounds
HA : Humic acids
FA : Fulvic Acids
CP : Calorific Capacity by Mass
EIS : Electrochemical Impedance Spectroscopy
CPE : constant phase element
HF : High frequency
LF : Low frequency line
RP1 : the polarisation resistance of the electrodeposition
Bavg: Average value of the magnetic field strength
FMI : Magnetic force intensity
LFWA : Magnetic field strength
CSH : calcium silicate hydrate

SEM	:	Scanning Electron Microscopy
SEM	:	Scanning Electron Microscope
C-S-H	:	calcium silicate hydrate

Summary

A new technology, based on the physical treatment of water by an electromagnetic field, can be a solution for the valorisation of water resources and offer new horizons for human technology, in synergy with its environment. The present study aims to evaluate and understand the effects of electromagnetic technology. To this end, laboratory experiments under controlled conditions were carried out to understand the mechanisms by which water treated with the electromagnetic field affects the physico-chemical properties of the water. It has been claimed that passing water through an electromagnetic field improves the chemical, physical properties and bacteriological quality of water in many different applications. Although this electromagnetic water treatment process has been used for decades, it is still in the realm of pseudoscience. If the claims of water treatment with electromagnets are confirmed, the process of magnetising water has gained enormous momentum worldwide. A large number of peer-reviewed journal articles have reported conflicting claims about this type of treatment. Some of the most beneficial applications of magnetically treated water include improving the mechanical strength of mortars and improving crop yields with reduced water use. Currently, there are many grey areas surrounding these new technologies and the number of question marks surrounding this topic is increasing. Many research papers are beginning to develop theories on the mechanism of the treatment. Most of the experiments carried out on this promising topic have revealed the role of electromagnetically treated water in improving the physical and chemical properties of water. It is in this context that this work has seen the days whose main objectives are to establish relevant experimental protocols able to offer a better understanding of the phenomenon of magnetisation of water and its effects on the physical and chemical properties of water. The results presented in this manuscript will serve as a basic reference for future works that investigate new perspectives offered by the use of the electromagnetic field especially in chemistry, biology and civil engineering... .

GENERAL INTRODUCTION

I. LITERATURE REVIEW

I.1 Introduction

The first question that arises on the subject of water magnetisation is, what does magnetised water mean? Magnetised water is water that has been magnetised by passing through a magnetic field. It is a new physical treatment of water, inexpensive and very environmentally friendly. The cost of this new technology is very low, as it requires only small installations. The effects of magnetism on water have been the subject of much controversy. Several works claim that magnetised water gives remarkable performance in reducing limescale (Alim et al 2006), increasing crop yields (Lin and Yotvat, 1990), health benefits (Yue et al. 1983), changing pH (Busche, 1985), reducing water stress (Cho & Lee, 2005) and increasing compressive and tensile strength of cement (Nan et al. 2000).

On the other hand, other scientific journals and researches claim that water magnetisation has no effect, and that the current successes could not be replicated (Krauter et al. 1996). Currently, there are hundreds of published papers and experiments on magnetic water treatment with a substantial percentage of success in treatment. The properties of water remain difficult to pin down, as it carries a variety of foreign bodies. It contains particles in the form of micro-contaminants and other dissolved solids, which adds to the confusion about magnetised water. Many scientists claim that certain chemicals in water determine the success rates of treatment with electromagnetic fields. All over the world, in different laboratories, the properties of water treated by the electromagnetic field vary from one experiment to another, and this is due to the mineral composition dissolved in the water, hence the need to distil the water before any electromagnetic treatment.

I.1.1 Aims and objectives

The present work investigated carefully the impact of electromagnetic fields on the physico-chemical properties of water, namely: pH, conductivity, density, TDS, temperature, transmission, absorption, turbidity, surface tension and heat capacity. In particular in the field of irrigation. In fact, the applications for which the effect of magnetised water was tested are the following.

- Germination of lettuce
- the growth of lettuce
- Formulation of fly ash and hearth mortars
- Optical properties

- Formulation of organic fertilizers
- Electrical properties
- Thermal properties
- Electromagnetic memory of water

This research work was carried out based on our expertise and on the literature. The aim of this research work is to offer the general public a manuscript that can serve as a reference for people who want to get involved in the use of magnetised water.

I.1.2 Scope

This thesis project aims to conduct experiments to obtain a

Broad understanding of the effect of the electromagnetic field on the physico-chemical properties of water.

The water used in the experiments includes:

- pH of tap water
- Formulation of tap water mortars
- Thermal study of tap water
- Optical study of tap water and distilled water
- Germination of lettuce using tap water and distilled water
- Growth of lettuce by irrigating with tap and distilled water
- Electromagnetic memory of tap water
- Electrical study of tap water
- Formulation of an organic fertilizer using tap water
- Study of the physico-chemical parameters of tap water

I.1. 3 Overview of the thesis

The first part will consist of a review of the literature and previous research on the subject. The second part will provide a detailed explanation of the methods and materials used. The third part consists of the in-depth results and discussion obtained. The last part will provide a conclusion of the work carried out and will give perspectives in this field.

I. 2 Review of the literature

I.2.1 History of magnetised water

Magnetic technology has been discovered since 1803 (Brower, 2005). Indeed, it all started with soup kettles. These kettles were placed on a fire with large stones at the bottom to prevent them from being blown around by the wind. These kettles were all made from the same metal. Yet two of them did not form hard scale. Instead, they had a soft, powdery substance that could be easily removed. It was later found that these two pots were stabilised by magnetite stones, which are composed of iron oxide Fe_3O_4 (Gholizadeh et al. 2008; McMahon, 2009; Bourget, 2011). These stones are widely found throughout the world, particularly in the Magnesia region, which is now modern Turkey (Cullity et al. 2011). Michael Faraday was the first researcher known for his fundamental work in the field of electromagnetism and electrochemistry. He reported that when an ionised fluid is subjected to a magnetic field, it generates an electric current (Vallée, 2004). From 1890 onwards, the magneto-treatment of water became a controversial subject, with some treating it as a "gimmick" that could not be considered a scientific object (Marshutz, 1996).

In the early 1900s, it was no longer alchemists who travelled the world in search of the Fountain of Youth. But scientists and researchers were exploring the question of longevity. In fact, there are five places on earth where people regularly live longer than 120 years. And they do so in good health, virtually free of cancer and tooth decay, remaining robust and strong, and women are able to have children at an advanced age. The best known population is the one living in the Hunza Valley in northern Pakistan. They live between 120 and 140 years. Dr. Henri Canda, from Romania, who is the father of fluid dynamics and won the Nobel Prize at the age of 78, spent six decades studying the water of the Hunza to find out what it was about this water that was so beneficial to the body. He found that the water did indeed have strange properties. Its boiling and freezing temperatures, viscosity and surface tension did not match the usual measurements. Subsequently, Dr. Patrick Flanagan confirmed these anomalous properties and tried to recreate them by synthesis. After years of study, Flanagan was able to modify certain physico-chemical properties using various techniques (electromagnetic and magnetic fields etc.), however these qualities were only transient, so the change is not permanent. Magnetic technology was first used in industry in the 19th century. The American researcher Porter was the first to propose a non-chemical device to control deposits in water boilers in 1865. The first magnetic field water purification device was patented in the United States in 1873. In 1890, Faunce and Cabell invented an electromagnetic device to treat boiler water (Ambashta et al. 2010). In 1936, a company called "Solavite", founded in France, started marketing a MTD (Magnetic treatment device). In 1954, other companies manufactured water conditioners based on a magnetic field (Marshutz, 1996). In 1960, Krylov and Tarakonova were the first to assert the auxin-like beneficial effects on germination observed following the exposure of seeds to a magnetic field (Maffei, 2014). In 1963, Boe and Salunkhe showed that the magnetic field also acts on the ripening of tomato fruits (Maffei, 2014). In the 1990s, many credible research institutions took on the task of evaluating the results of this treatment. Today, there are many varieties of BAT, with selling prices ranging from $100 to over $10,000 (Otsuka et al. 2006). However, the controversial debate on the effectiveness of magnetically treated water persists to this day.

I .3. Definition

Static magnetic field water treatment is defined by Ali et al. (2014) as a simple and efficient physical treatment, where water flows through a magnetic field or combination of magnetic fields that influences some of its physico-chemical properties, without changing the chemical structure. This treatment requires the installation of a device in the main water channel, where the water is treated after passing through it. These devices contain static magnets, usually of the NdFeB type, which create specific strong magnetic inductions. The action of this induction (called magnetisation) is explained by the relative movement of electric charges, which is the only property that purifies water independently, influencing its physicochemical properties (Ambashta et al., 2010). These devices are called magnetrons or magnetisers. There are two types: devices mounted on the water pipe designed to be wrapped around the pipe, and others mounted in the water pipe that require the removal and replacement of part of the pipe by the device (Ali et al., 2014).

I.4 Effect of magnetic treatment on water properties

Water is an indispensable element for our life. It is necessary for the growth and development of humans, animals and plants. The water molecule is polar and because of this polarity, the molecules bond and release with each other, 1012 times per second, with the electronegative oxygen atoms attracting the electropositive hydrogen atoms (Teixeira, 1999). This bonding is called hydrogen bonding. These bonds are particularly found in ice and in some organised forms of water. In general, each group or cluster, called "clusters", contains about 100 water molecules at room temperature and come in a wide variety of shapes and sizes (Su et al. 2003). This important cohesion and strong link between water molecules are at the origin of its abnormally high physical and chemical properties (maximum density at 3.98°C, high surface tension at 72.7 mN/m at 20°C, viscosity of 8.9.103 P, electric dipole moment μ=1.83 D, relative permittivity at 25°C: εr= 78.3, ...) (Vallée, 2004). These properties allow it to act as a solvent, reagent, temperature stabiliser and molecule with cohesive properties (Ibrahim, 2006). It would therefore be interesting to study the effect of the action of external factors on the molecular structures and changes in the properties of water. Smirnov (2003) observed that water can receive signals from magnetic forces as a result of the rotational movement of the proton spin. Therefore, when water is subjected to a magnetic field, its physico-chemical properties change, both on the microscopic and macroscopic scales (Pang et al. 2008b; Hao et al. 2012). Incidentally, it becomes more energetic and biologically active (Yavuz et al. 2000; Tai et al. 2008, 2014).

I.4.1. Microscopic scale

From a microscopic point of view, magnetic treatment creates changes in the physico-chemical properties of water. These changes affect the electronic, atomic and molecular structures, but the chemical constitution remains unchanged (Deng et al. 2007; Pang et al. 2008a, 2008b). In fact, the application of a magnetic field improves the polarity of the water molecule, influences the size of the distributions and arrangements of particles in water and affects the transition of valence and inner layer electrons (Lebkowska, 1991; Krzemieniewski et al. 2002). The measurement of the X-ray diffraction spectrum shows changes in the distribution of electrons, due to electron movements (Pang et al. 2008b). This result is in agreement with those of Lee et al. (2013) who suggested that the magnetic field influences the proton spin, promoting proton transfer in the closed chains of hydrogen bonds (Madsen, 1995; Hosoda et al. 2004; Lee et al. 2013). On the other hand, it has been reported that in the presence of a magnetic field, the magnetic force breaks the water clusters and reduces them to small chains (Yu et al. 1998; Su et al. 2003). As a result, the activity of water is enhanced, which can be explained by the formation of new hydrogen bonds (Xantheas, 2000). Pang et al. (2008a, 2008b) and Moosavi et al. (2014) confirmed these results by reporting an increase in the clustering of water molecules due to the increase in the strength of hydrogen bonds. Chang et al. (2006) found that the number of hydrogen bonds increased by about 0.34% when the magnetic field strength was increased from 1 to 10 T. Toledo et al (2008) explained that the magnetic treatment influences the hydrogen bonding network, weakening the intra-cluster bonds, breaking them and thus forming smaller clusters with stronger inter-cluster bonds. This confirms the hypothesis of Zhou et al. (2000), that the magnetic field would weaken or even break the hydrogen bonds, resulting in the formation of a large number of monomers of water molecules. In addition, it has been reported that the magnetic field rearranges the dipoles of water molecules which improves the organisation of clusters in water (Jeon et al. 2001; Lee et al. 2009). Lee et al. (2013) confirmed this structuring of water molecules by measuring the spin-spin relaxation time, which can be used to determine the degree of organisation of water molecules. Moosavi et al (2014) linked this perfect alignment of water molecules to the change in charge distribution following exposure of water to a magnetic field. Brower (2005) reported an increase in the tetrahedrality of the water molecule, adding that for the H_2O molecule, the bonds of the hydrogen and oxygen atoms are changed from a triangular structure to a linear structure.

I.4.2. Macroscopic scale

Surface tension is related to molecular interactions because it results from the increase in cohesive energy. It is, therefore, an important physico-chemical indicator of the change in the structure of water following treatment with a magnetic field (Cai et al. 2009). The literature reports that the surface tension force decreases after exposure of water to a magnetic field (Amiri et al. 2006; Pang et al. 2008b). This decrease results in a decrease in molecular energy. Cai et al. (2009) deduced that water becomes more stable, which supports the hypothesis that new hydrogen bonds are formed during magnetic treatment processes. The magnetic field thus increases the hydrophobicity of water due to the attenuation of the contact angle of magnetically treated water (Pang et al. 2008b). This is explained by changes in the size distributions of molecules in water (Hołysz et al. 2002; Chibowski et al. 2003; Cho et al. 2005; Amiri et al. 2006). Otsuka et al. (2006) added that the attenuation of the contact angle is conditioned by the presence of O_2 in the water. The changes observed for electrical conductivity are controversial. Some authors have reported an increase in electrical conductivity (Ranade, 1990; Ponomarev et al. 1999; Jeon et al. 2001; Pang et al. 2008b; Lee et al. 2009). Pang et al. (2008b) explained this result by the increase in the number of charged particles, which forces the polarised water molecules in the molecular chains or group of molecules to separate as H^+ or $H_3O^+$and OH^-. Other authors, on the other hand, reported a decrease in electrical conductivity (Akopian et al. 2004; Kronenberg, 2005). Kronenberg (2005) explained his hypothesis by the fact that magnetically treated water contains fine colloidal molecules (in a state of constant motion resembling Brownian motion) and electrolytic substances that respond to magnetic treatment by their increasing capacity to sediment. These challenges have been justified by the type of magnetic treatment, the chemical quality of the water used and its purity (Vallée, 2004; Lee et al. 2013). On the other hand, the refractive index of magnetically treated water increases (Pang et al. 2008b), so the dielectric constant increases after magnetic treatment (Hosoda et al. 2004; Holysz et al. 2007). Indeed, hydrogen bonds become more stable in the presence of a magnetic field, therefore, optical properties such as ultraviolet absorption and refractive index increased (Hosoda et al., 2004). In addition, Bruns observed that the treatment of water with a magnetic field resulted in a 30% change in light absorption intensity (Bruns et al., 1966). The viscosity decreases as a result of the magnetic treatment, which is due to the formation of a large number of chains of molecules, resulting in an increase in the flow velocity. Therefore, the water becomes more wettable (Pang et al. 2008b; Tai et al. 2008). However, other authors have observed an increase in the viscosity of water during the magnetic treatment process (Cai et al. 2009). On the other hand, Tai et al. (2008) reported a decrease in the pH value from 9.2 to 8.5 after treating their water samples with a magnetic field. Indeed, the magnetic treatment increases the formation of hydroxyl ions and thus reduces the acidity of the water. So

That Quickenden et al (1971) mentioned no change.

I.4.3. Mechanism of action

The principle of the mechanism of action of the magnetic field on the properties of water remains a controversial subject that has not been scientifically well understood, although various hypotheses have been proposed. As previously reported, the treatment of water with a magnetic field significantly influences the zeta potential and size distributions of the molecules (Parsons et al. 1997). This can be attributed to the Lorentz force exerted on the moving ions or on solid charged particles, especially on the spin proton (Higashitani et al., 1993; Madsen, 1995). On the other hand, Cai et al. (2009) attributed the effects exerted by magnetic treatment to changes in the molecular energy of water. They explained that the change in molecular energy is an indicator of the reorientation, formation or breaking of hydrogen bonds. Thus, the inner structure of water becomes more stable and inactive. Thus, the activation energy increases and water will need more energy to produce a reaction. Cai et al (2009) added that the rotational movements of the water molecules become very slow, which can be explained by the strengthening of hydrogen bonds. They even simulated the action of the magnetic field on water as a result of the lowering of temperature. Despite the mechanisms proposed above, there is no scientific consensus on the mechanism of the effect of magnetic fields. On the other hand, some authors reported that the condition of the water persists, after the treatment is stopped, for up to 12 h (Fu et al. 1994; Wang et al. 1997).On the other hand, Lee et al. (2013) observed that the effects of magnetic treatment on the properties of the water disappeared after 6 h. However, Kney et al. (2006) noted an effect of

Memory that lasts up to 120 h, while other authors have reported that the effect observed would be perceptible for 200 h (Coey et al. 2000).

I.5 Summary of the literature review

For many years, the effects of magnetic fields on water have been of interest to physicists, chemists and biologists (Parsons et al. 1997). Nowadays, magnetic processing is of interest to several sectors such as medicine, environment, construction, industry, etc. In this work, we focus on the application of magnetic fields to irrigation water. For several decades, researchers have found that the process of treatment by a magnetic field independently contributes to water purification by influencing the physical and chemical properties of contaminants in water (Liu et al. 2011). Indeed, the ability of water to receive signals from magnetic forces allows it to directly influence living cells (Smirnov, 2003), hence the use of the magnetic field for water purification at the bacteriological level (Yadollahpour et al. 2014). In this context, Johan et al. (2004) reported that magnetic treatment is a promising process that can improve the separation of suspended particles in wastewater. Furthermore, Strašák et al. (2002) stated that the ability of bacteria to form colonies decreases with increasing magnetic field intensity and exposure time. Thus, the magnetic field has a "bactericidal" effect. Finally, Starmer (1996) reported a reduction in the growth of algae and fungi. Furthermore, magnetic treatment has been successfully implemented in the industrial sector to reduce heating and disinfection expenses, specific to swimming pool management (Starmer, 1996). Lipus et al (2013) observed an improvement in the whiteness of cotton after using magnetically treated water for washing, which consequently leads to a minimisation of detergent use. They also reported easier stain removal, low temperature disinfection and protection of fabric colour quality. Research in Japan, Europe and China reports that concrete wetted with magnetically treated water provides about 10% greater cohesion strength than ordinary concrete, and is completely waterproof. Therefore, it can save 5% of cement dosage, decrease concrete cracking and improve freezing resistance (Su et al., 2003).Selim (2008) reported that magnetically treated water has the ability to accelerate the drainage of salts, with almost double. Indeed, the process of treating water with a magnetic field can increase the movement of salt to the depths below the rhizosphere (Hilal et al. 2000; Mostafazadeh-Fard et al. 2011). Mohamed et al. (2013) reported that the percentage of salt removed after irrigation with magnetically treated water is more than 25%.This process of treating water with a magnetic field is also used in livestock farming entraining drinking water. Gholizadeh et al (2008) observed a decrease in mortality rate and an increase in feed conversion rate in poultry that consumed magnetically treated water. It has been reported that treatment of catfish growth media with a magnetic field has altered cell size and density as well as the size of the nucleus on catfish liver cells (Bourget, 2011).However, the two main benefits of magnetic water treatment are the reduction or elimination of calcium deposits and the improvement of crop yields by minimising water use efficiency (Stuyven et al., 2009; Yadollahpour et al., 2014).

MATERIALS AND METHODS

II.1 Electromagnetic device design

II.1.1 Definition

The electromagnetic device is a physical water treatment technology based on concepts from quantum physics and electrodynamics. The water passes under the effect of the electromagnetic field in continuous flow through a cylindrical pipe (**Fig.1**). The experimental apparatus consists of a pump immersed in a tank full of water. The pump is connected in such a way that the water can flow through the region where the electromagnetic field exists. In the other tank, the magnetised water is collected. This experiment is carried out cyclically (**Fig.1**).

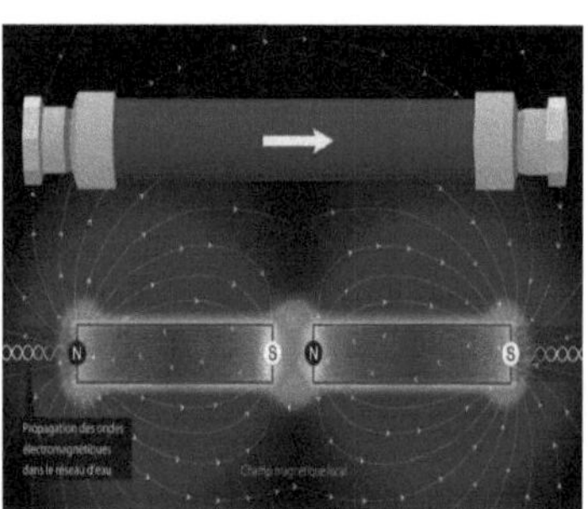

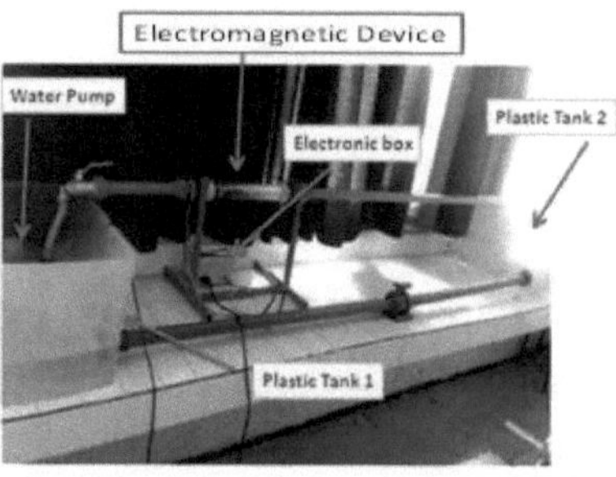

Fig. 1 Electromagnetic field generator equipment

II.1.2 Characterisation of the electromagnetic device in vacuum

In order to know the currents flowing in the two coils of the device and the frequency of the electromagnetic field, the characterisation of the electromagnetic device was carried out under vacuum according to the experimental protocol illustrated in **Fig. 2** and **Fig. 3**.

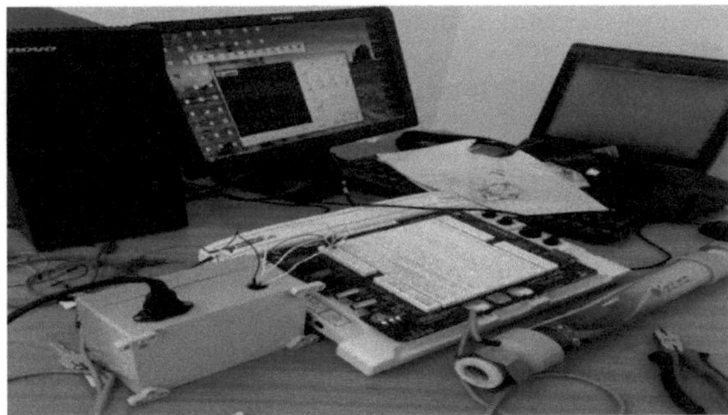

Fig. 2 Measuring the current flowing in the coils

Fig. 3 shows that the signal shape is square, with a frequency of 6.116 kHz and a peak-to-peak value of 10.54 V in one cycle and 3 V in the other cycle (this difference is due to hysteresis).

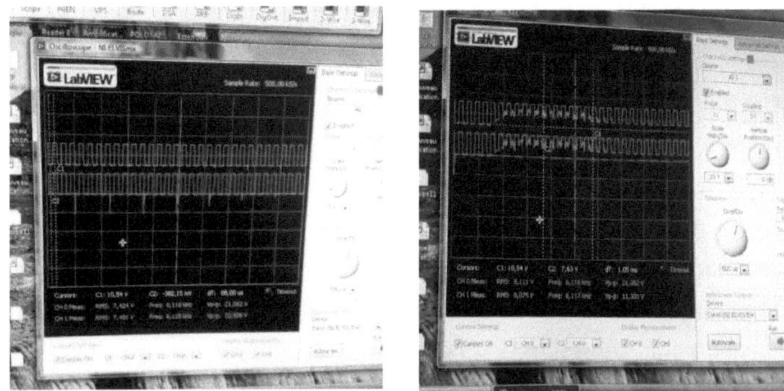

Fig. 3 Visualization of the signal shape. Blue: voltage across coil 1, green: voltage across coil 2

II.2 Study of changes in the properties of magnetised water and its applications

II.2.1 Experimental measurement of pH

The pH of the tap water in the tank was measured before the water was treated with the electromagnetic field, after the water was circulated inside the magnetising device at two flow rates 0.18 l/s and 0.6 l/s for 5 min, 10 min and 20 min. Then the pH of the magnetised water was measured during a time interval of 120 minutes.

II.2.2 Assessment of thermal and surface tension behaviour

II.2.2.1 Heating temperature measurements

The study and analysis of water was carried out at the Laboratory of Condensed Matter Physics, Faculty of Science, Chouaib Doukkali University, Eljadida Morocco. The difference between the evaporated volume of magnetised water ΔVe (MW) and the evaporated volume of non-magnetised water ΔVe (NMW) at a fixed temperature is ΔV, which is defined by equation 1 (**Eq. 1**).

$$\Delta V = \Delta V_e(MW) - \Delta V_e(NMW) \quad \textbf{(Eq. 1)}$$

The difference between the evaporated mass of magnetised water Δm (MW) and the evaporated mass of non-magnetised water Δm (NMW) at a fixed temperature is Δm given by Equation 2 (**Eq. 2**).

$$\Delta m = \Delta m(MW) - \Delta m(NMW) \quad \textbf{(Eq. 2)}$$

Measurements of the heating temperature over time of tap water and magnetised water range from 5 to 120 min. Previously, several research projects were conducted on the influence of the static magnetic field on water, with opposite results. Thus, to evaluate the influence of the electromagnetic field created by the electromagnetic device on the heating rate of water, experiments were carried out with the instruments shown in (**Fig. 4**).

Fig. 4 Measurement of the water heating rate.

Figure 4 shows the following experimental components.

- A hot plate supplied by S.B.S INSTRUMENTS, S.A.
- A thermocouple for temperature measurement supplied by PHYWE.
- A digital stopwatch from HANHART.

- Two 200 ml PYREX graduated beakers.

Experiments were carried out to study the evolution of water temperature over time, to confirm that the temperature of (MW) can increase more rapidly than tap water (NMW), and the heat transfer rate of (NMW) was compared to that of treated (MW) water. This method is based on heating water in a beaker using a hot plate, a digital thermocouple and a stopwatch (**Fig. 5**).

Fig. 5 Calorimetric measurement of magnetic water (MW) and tap water (NMW)

Next, 200 ml of tap water (T= 26.5°C) was poured into the beaker, which was placed on the hot plate set at 150°C. Simultaneously, the stopwatch was started and the temperature was measured at different time intervals in order to draw a curve representing the evolution of the water temperature as a function of time.

II.2.2.2 Cooling temperature measurements

This study tracks changes in water temperature over time to determine whether magnetised water, generated by electromagnetic device technology, is colder than ordinary water. The aim of this study is to compare the cooling rate of tap water (control) with that of magnetised water. The method initially consists of heating the water in a beaker with a hot plate until the temperature of the water sample reaches 80 °C, and then the timer was started (**Fig. 6**). By measuring the temperature with a thermometer at varying time intervals, curves can be drawn showing the temperature of the water as a function of time.

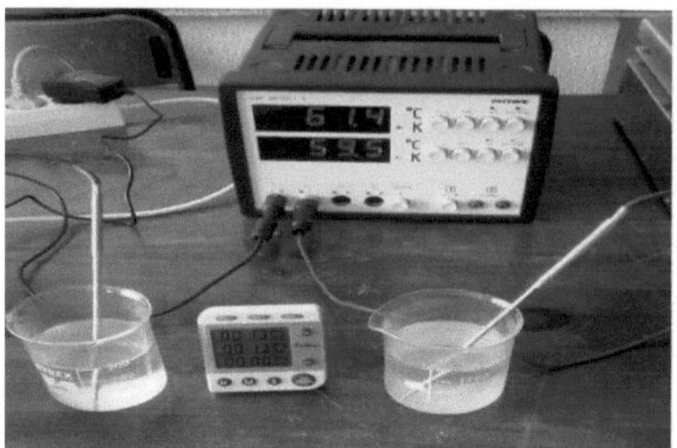

Fig.6 Measuring the cooling rate of tap water and magnetised water

II.2.2.3 Surface tension measurement

Surface tension is a resulting cohesive force that minimises the number of molecules on the surface of a liquid. This creates a kind of invisible shell that occupies the smallest possible surface area. The surface tension represents the film strength of the liquid surface (Cefalas et al.2010).

. The measurement of surface tension consists of weighing a drop falling from a capillary of known radius. To a first approximation, the forces that apply to the drop are its weight $P = mg$, and the force related to the surface tension γ at the capillary $F = 2\pi r\gamma$.

At a specific moment when the drop is detached, the weight of the drop is equal to the capillary force $P = F$. This implies the application of Tate's law.

$m = 2\gamma\pi r/g$ **(Eq.3)**

Knowing the mass of the drop, the surface tension γ can be estimated by the following expression:

$\gamma = mg / 2\pi r$ **(Eq.4)**

To measure the surface resistance of all samples, the falling drop method was used with a pipette. The principle of the measurement is based on counting the number of drops corresponding to a given volume of water of density d at room temperature. The experimental measurements were carried out by the following instruments (**Fig.7**):

- A 100 ml PYREX beaker

A support

- Precision scale (OHAUS)
- Burette 25 ml (QUALICOLOR)

The burette was attached to the stand and the beaker was placed underneath to collect the drops formed. The burette was previously washed and rinsed with distilled water. A certain volume of liquid was poured into the burette. The volume of liquid protruded a few centimetres beyond the top line of the apparatus. The counting of the number of drops started when the meniscus of the liquid slightly reached the top line. The flow speed of the liquid should allow time for the drops to form before falling. The velocity was adjusted using Mohr's forces. The volume above the top line was used and the number of drops of liquid was counted while avoiding air flow when counting the drops (Tijing et al.2010).

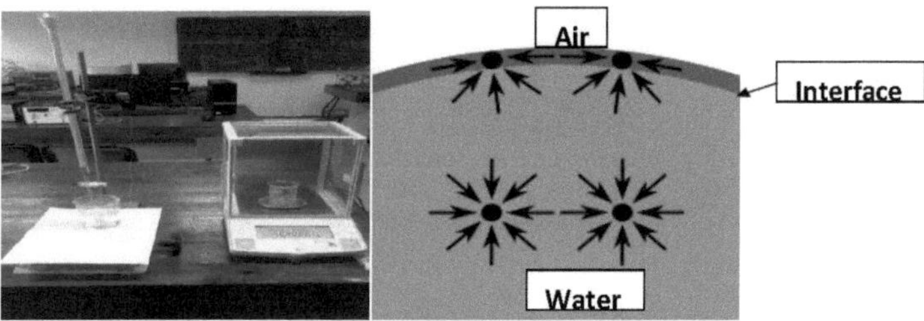

Fig. 7 Measuring the surface tension of magnetic water by the falling drop method.

II.2.3 Study of magnetised water by electrochemical means and by impudence spectroscopy.

II.2.3 .1 Electrochemical experimental methods

Electrochemical measurements were performed in a conventional three-electrode cell, using the Volta lab 40 potentiostat-galvanostat (Tacussel-Radiometer PGZ301), monitored by the Volta master 4 option analysis software model at steady state. Impedance patterns were recorded over a frequency range of 100 kHz to 10 MHz with 10 points per decade and an amplitude of 10 mV (peak-to-peak) of the excitation signal.

II.2.4 Study of some optical properties of magnetised water

Fourier transform infrared spectroscopy (FTIR) was used to measure the optical properties of distilled, tap and magnetised water. For comparison, the transmittance of electromagnetic field (EMF) treated and untreated samples were all measured. The FTIR measurements were carried out on the technical platform of the Faculty of Sciences of El Jadida with a Fourier Transform Infrared Spectrometer (Thermo Scientific), equipped with a temperature controlled semiconductor diode laser, a KBr optimised beam splitter from 7800 to 350 cm-1 optimised in the mid-infrared. An electrically cooled deuterated triglycine sulphate (TEC) standard detector with fast recovery (DTGS) for maximum linearity of response, a background reference was taken with the empty cell before measuring a series of samples. The results obtained are analysed by the OMNIC cloud-based FTIR software shown in **Fig. 8**.

Fig. 8 Fourier transform infrared spectrometer

II.2.4 Electromagnetic water memory

II.2.5 .1 pH measurement

The pH is measured with an Adwa AD1000 pH meter as shown in **Fig. 9**.

Fig. 9 Measuring the pH meter of magnetised water

II.2.5.2 Conductivity measurement

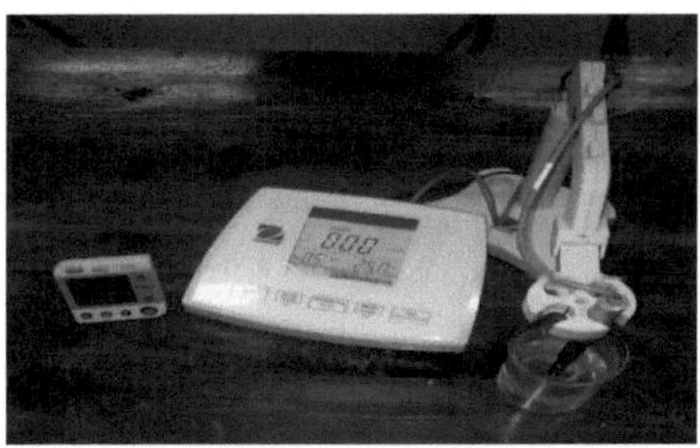

Fig. 10 Conductivity, salinity and TDS measurement of magnetic water by conductivity

II.2.5.3 Density measurement

The magnetic water was placed in a 250 ml volumetric flask. Then its mass was measured using a sensitive EXPLORER® PRECISION balance as shown in **Fig. 11**.

Fig. 11 Measuring the magnetic water mass with a balance

The density of magnetic water can be defined as $\rho = mV$ where, m is the mass and v is the volume

II.2.5.4 Temperature measurement

The temperature of the magnetised water is measured by a PHYWE digital thermocouple marker as shown in **Fig. 12**, which consists of a pair of different metals. If the two contact points have different temperatures, the difference between the contact potentials can be measured as the thermal potential.

Fig. 12 Measuring the magnetic water temperature

II.2.5.5 Measuring transmission, absorbance and turbidity

MULTITESTS is a multi-parameter photometer as shown in **Fig. 13**, manufactured by AQUALABO GROUP for the analysis of more than 40 parameters (transmittance, absorbance, turbidity, nitrate nitrogen, ammonia nitrogen, phosphate tetraoxide, iron, nickel, chemical oxygen demand (COD) ..... Compatible with liquid reagents such as water.

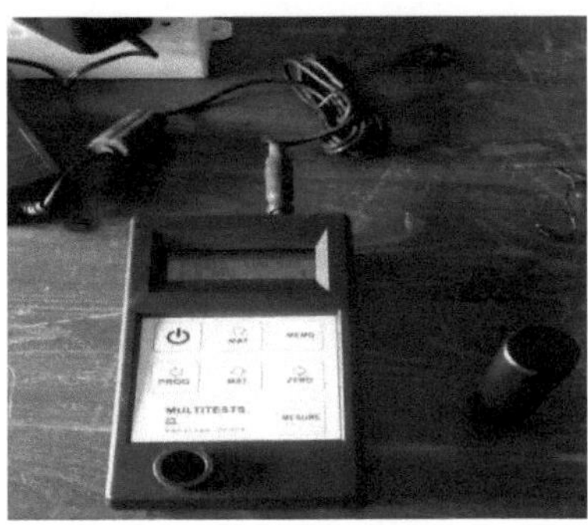

Fig. 13 Spectrophotometer for measuring transmission, absorbance and turbidity of magnetic water

II.2.5 Parameters influencing the magnetisation process of water

The aim of this work is to study the parameters influenced by the magnetisation process. In other words, the characterisation of the parameters that influence the magnetisation process of tap water has been studied. The magnetic device generates an electromagnetic field acting as an external pulse, which will influence the magnetisation of the tap water. The circulation of tap water (TW) under the effect of the electromagnetic field (EMF) generates magnetised water (MW). The parameters were estimated after the electromagnetic field was switched off. The magnetisation time of the water was controlled by the kobra interface using a Latis pro software. **Figure 14** shows the method of heating the magnetised water and tap water to perform a comparison study by visualising the equilibrium electromagnetic field during heating and cooling of both types of water. The electromagnetic field monitoring was performed by the kobra software. The data was recorded automatically **Fig. 15**.

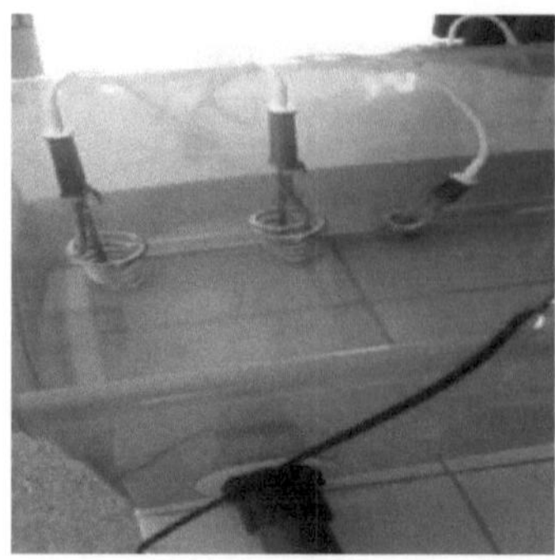

Fig. 14 Method of heating magnetised water.

Fig. 15 Measurement of the EMF intensity of magnetised water as a function of time.

II.2.6 Physico-chemical parameters of tap and magnetised water

The physical and chemical analysis of the tap and magnetised water gave the following results (**Table 1**).

Table 1. Physico-chemical analysis of tap water and magnetised water

Parameters	Tap water	Magnetised water
pH	7.69	8.25
Conductivity (μS/Cm)	1695	971
TDS (mg/L)	848	485
Conductivity (PSU)	0.85	0.48
Calcium (mg/L)	78.5	79
Oxidability (mg/L)	0.41	0.49
Chloride (mg/L)	88.5	89
Temperature (°C)	17.8	17.8
Magnesium (mg/L)	23.83	23.9
Alkalinity (°F)	10.35	10.4
water content (meq/l)	2.94	2.98
Ammonium (mg/L)	0	0
Iron (mg/L)	0	0

II.3 Experiments related to magnetised water applications

II.3.1 Germination measurement of lettuce (Lactuca sativa L.)

Germination experiments were carried out at the Laboratory of Condensed Matter Physics of the Faculty of Science, El-Jadida, Morocco to study the effect of electromagnetic fields generated by the magnetic device on the germination of lettuce seeds with tap water. The test was carried out at an ambient temperature of 26 ± 2°C. Seeds of uniform size and shape without visible defects are used. The tests are performed with three types of water, DW: distilled water, TW: tap water and MW: magnetised water with different magnetisation time (tm). In this work, the electromagnetic device is mounted on an experimental system stand while water is circulated in a closed loop for 20min, 40min, 60min and 80min to obtain water magnetised by an electromagnetic field (**Fig. 1**).

II.3.1.1 Seed germination

Eighteen Petri dishes should be used for each test (three treated with MW tm = 80min, three treated with MW tm = 60min, three treated with MW tm = 40min, three treated with MW tm = 20 min, three treated with tap water and three treated with distilled water), each containing ten lettuce seeds for each test for the estimated germination period of four days (**Fig. 16**)

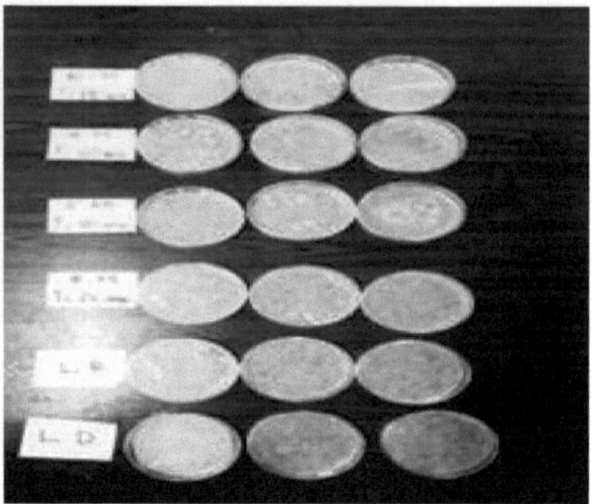

Fig.16 Effect of electromagnetic fields on the germination of lettuce seeds for each type of treatment.

For a statistical study, we used twelve Petri dishes for each test (4 Petri dishes treated with magnetised water, 4 with tap water and 4 with distilled water), each containing ten lettuce seeds for each test. Each test during the germination period is estimated to take four days.

II.3.2.1 *Measuring TDS of magnetic water*

The tap water was magnetised for 80 min in kinetic mode, and then germination monitoring of lettuce irrigated with this magnetised water was carried out in order to study the effect of electromagnetic field treated tap water (generated by the Aqua4 D device) on the growth parameters of lettuce plants. Lettuce seeds (Lactuca sativa L., variety: Lettuce Sucrine) of uniform size and shape without visible defects were used. Ten lettuce seeds were sown in Petri dishes containing blotting paper for further treatments.

The test was carried out at an ambient temperature of 26 ± 2ºC.Eighteen Petri dishes were divided into six main treatment types where the blotting paper was sprayed with a different type of water:

- DW: the blotting paper is sprinkled with distilled water;
- TW: the blotting paper is sprayed with normal tap water;
- MTW20: the blotting paper is sprayed with magnetically treated tap water for 20 minutes;
- MTW40: the blotting paper is sprayed with magnetically treated tap water for 40 minutes;
- MTW60: the blotting paper is sprayed with magnetically treated tap water for 60 min;
- MTW80: the blotting paper is sprayed with magnetically treated tap water for 80 minutes.

After 4 days of germination test, homogeneous plants were the target of the growth test and magnetised water was prepared in four different steps. In sterilised Petri dishes (three copies for each test) were soaked with 4 ml of distilled water, tap water or magnetised water, from which homogeneous seedlings were placed. Four plants (x3) treated with tap water, four plants (x3) treated with distilled water and four plants (x3) treated with magnetised water (20min), four plants (x3) treated with magnetised water (40min), four plants (x3) treated with magnetised water (60 min), four plants (x3) treated with magnetised water (80 min). Petri dishes (distilled water, tap water, magnetised water) were grown for 4 days in a photoperiod of 16h / 8h and at a temperature of 23°C +/- 2°C. For each treatment (distilled, tap, magnetised water), 12 plants were used to measure root elongation and fresh weight. The plants were then transferred to an oven for 48 hours at 45°C to measure the weight of dry matter.

II.3.2.1 *Measuring TDS of magnetic water*

Soluble solids content (TDS) measurements were made using a conductometer (Fig. **10**). The tests were undertaken every day for seven days.

II.3.2.2 Measurement of morphological parameters of lettuce (Lactuca sativa L.)

For a statistical study, twelve Petri dishes were used for each test (four treated with magnetised water, four with tap water and four with distilled water), each containing ten lettuce seeds for each test. Each test was carried out during the estimated germination period of four days (Fig. 17).

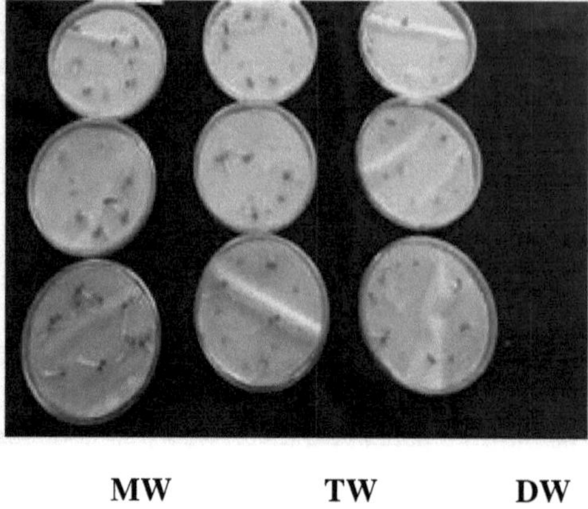

MW **TW** **DW**

Fig. 17 Effect of magnetic water on the growth of lettuce seeds for each treatment type

II.3.2.3 Experimental measurement of root elongation in lettuce

Root elongation was measured with a millimetre paper, calculating the average elongation of the plants in the different treatments by the following method:

Average root elongation of lettuce (cm) = (the total sum of root elongation of plants) / (the total number of plants)

II.3.2.4 *Measurement of* dry and fresh mass of lettuce

At the end of the experiment, the roots and aerial parts of the lettuce were collected in aluminium foil, and the fresh mass in (mg) was calculated with precision balance (**Fig. 18**). Then, the lettuce plants were placed in an electric oven (**Fig. 19**) to dry at a temperature of 45°C for 48 hours until a constant mass was reached, after which the lettuce plants were measured again to calculate their dry mass in (mg).

Average fresh lettuce mass (mg) = (The total sum of fresh lettuce mass) / (number of fresh lettuce plants)

Average dry mass of lettuce (mg) = (The total sum of dry mass of lettuce) / (number of dry lettuce plants)

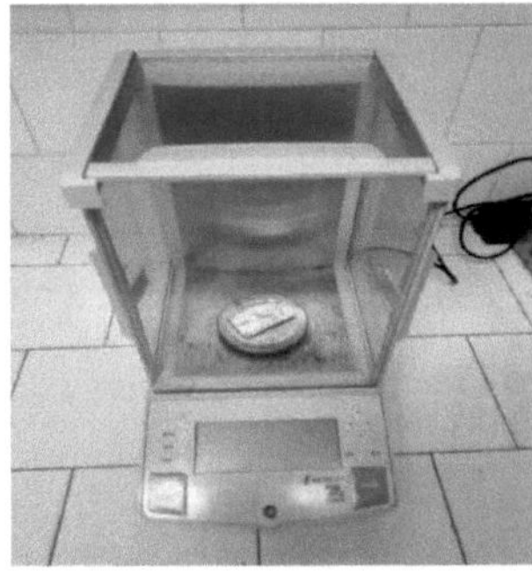

Fig. 18 Measurement of dry and fresh weight of lettuce plants

Fig. 19 Drying lettuce in an electric oven

II.3.2 Mortar formulation and characterisation

a. Preparation of the mortar

Mortar is a composite material, essentially consisting of a mixture of three ingredients: sand, hydraulic binder (cement or lime) and water, which is usually tap water. In this study, fly ash (FA) is valorised in cement mortar by replacing it with different percentages of (FA) with FA/cement mass ratios equal to: 10%. In order to determine the mechanical properties, the mortars are prepared according to the European standard NF EN 196-1 which requires that the normal mortar is composed of, by mass, one portion of cement (450g), three portions of sand (1350g) and half a portion of water (225g), i.e. with a water/cement ratio: W/C = 0.5. The products were mixed for 4 minutes according to the requirements of the standard in a 5 L mixer at room temperature and relative humidity above 50% (Helmuth et al.1987). Before starting the mechanical study, the mortars were prepared as prismatic specimens of dimensions 4 × 4 × 16 cm as shown in **Figure 20**. Thus formed these specimens were stored in the open air and then tested after 28 days in terms of mechanical strength in bending and compression.

Fig. 20 Fly ash mortar samples

Table 2 shows the complete formulation of the mortars. The B0 mortar is the normal mortar which was taken as a reference. The B10 mortar was obtained by replacing part of the cement with 10% fly ash by mass.

Table 2. Formulation of mortars with 10% fly ash

Elements	Fly ash(g)	Cement (g)	Sand (g)	Water (g)
B10	45	405	1350	225

II.3.3 .2.1 Analysis of materials to be used

II.3.3 .2.2 Cement

The cement used in the preparation of mortars is CEM II (CPJ 45) defined in the Moroccan standard NM 10.1.004. It is a Portland cement composed of 81.5% clinker, 12% limestone and the rest is gypsum which regulates the setting. The chemical composition of CPJ 45 cement is presented in **Table 3** below:

Table 3. Chemical composition of CPJ45 cement

Cement	% SiO_2	% Al_2O_3	Fe_2O_3	CaO	MgO	% SO_3	K_2O
CPJ 45	17	5	3	63	2.3	3.33	1.2

II.3 .3 .2.3 Sand

The sand used in the mixtures came from the coast (Oualidia), its particle size analysis was carried out according to the NF P18-560 standard [Zalaghi et al.2018] with an electric sieve. The results obtained are summarised in the graph in **Figure 22**.

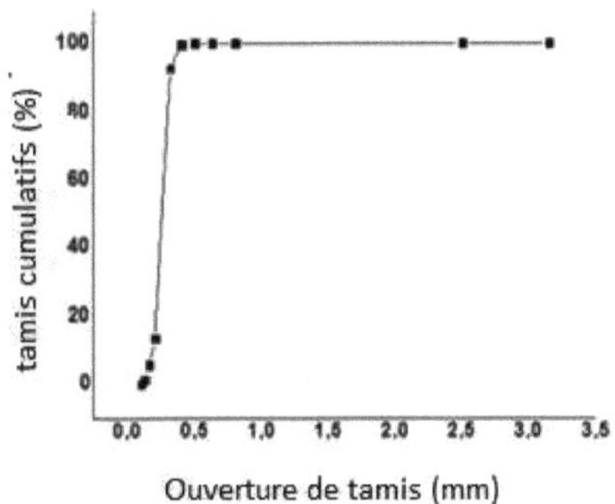

Fig. 21 Sand grading curve

According to the grading curve, the corresponding Cu uniformity coefficient is 1.7 and the predominant grain size is between 0.2 mm and 0.315 mm. Therefore, the sand used is uniform and composed mainly of fine grains. The bulk density of the sand used was determined by considering a pore (void) proportion of 41.5% according to the reference (Fan et al. 2005). The summarised results are presented in **Table 4**.

Table 4. Density data

Bulk density withc compaction	Bulk density with compaction	Actual density
1.43 g/cm3	1.54g/cm3	2.77g/cm3

The actual density obtained, which is 2.77 g/cm3, corresponds well to the specific gravity of fine sand, which generally varies around 2.65.

II.3.3 .2.4 fly ash

Coal fly ash is a fine powder resulting from the combustion of pulverised coal in the boilers of the Jorf Lasfar "JLEC" thermal power stations. This ash is carried along by the gaseous discharge flows generated by the combustion reaction and is thus captured by electrostatic filters which separate it from the gaseous discharge. The fly ash samples studied are grey in colour and have a smooth, shiny surface (Moufti et al. 2003). They are fine and look like cement powder - see **Figure 22**.

Fig. 22 Visual appearance of fly ash

The fly ash collected was obtained by burning South African coal. Each sample was obtained by splitting several 5 kg samples. The splitting consists of dividing the sample into four equal parts. Two opposite parts are recovered and homogenised. One of the latter two parts is reframed and so on. The operation can be repeated three or four times to obtain a representative sample (Adamiec et al. 2005). **Table 5** shows the results for the chemical elements in fly ash.

Table 5. Chemical elements in fly ash.

% SiO_2	% Al_2O_3	Fe_2O_3	%∑ SiO_2 + Al_2O_3+Fe_2O_3		CaO	MgO	% SO_3	K_2O

57	34	3,4	94,4		5,05	0,02	0,5	0,03

The chemical composition of the ash shows that the sum of the percentages of the elements SiO2, Al2O3 and Fe2O3 is 94.4%, which allows it to be classified as aluminous silica ash. The mineralogical composition of the fly ash is shown in **Figure 23**. Two peaks of mullite and quartz were detected in this experiment. During combustion, these minerals change their structure and give rise to a weak crystallised part in the form of mullite and quartz and an amorphous part (Zalaghi et al.2018). This result is in good agreement with the mineralogical composition of bottom ash.

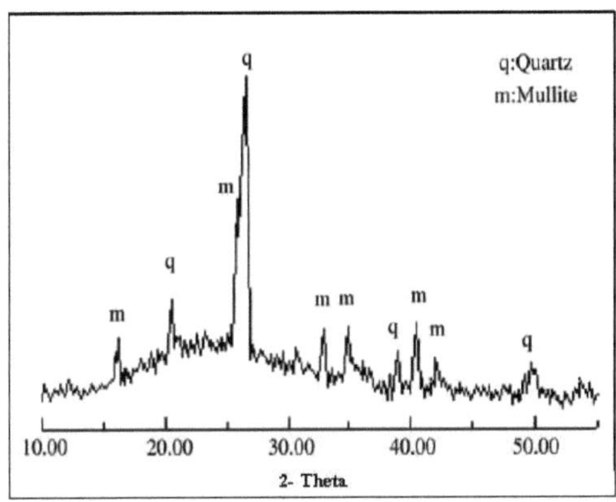

Fig. 23 Mineralogical spectrum of the fly ash used

II.3.3.2 Mechanical study: method

The measurement of the mechanical flexural strength of mortars was carried out on a material characterisation machine equipped with a three-point bending device, shown in **Figure 24** (Graich et al.2020). The experimental conditions used are described in EN 196-1. The specimen was placed symmetrically at the ends of the support so that the direction of the force was perpendicular to the longitudinal axis of the specimen. A pre-load was applied to avoid any sample play and to balance the displacement measuring device (**Fig. 24**).

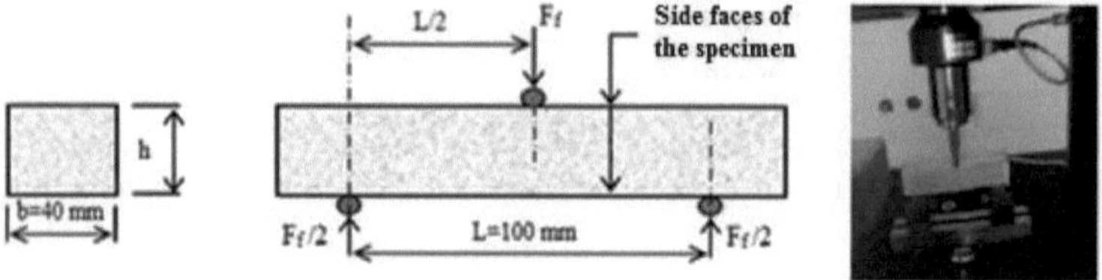

Fig. 24 Bending strength test device

Rf = 3FfL/2bh2 (**Eq**1)

After bending, each half-prism was reused to measure the compressive strength - see **Figure 25** (Graich et al.2020).

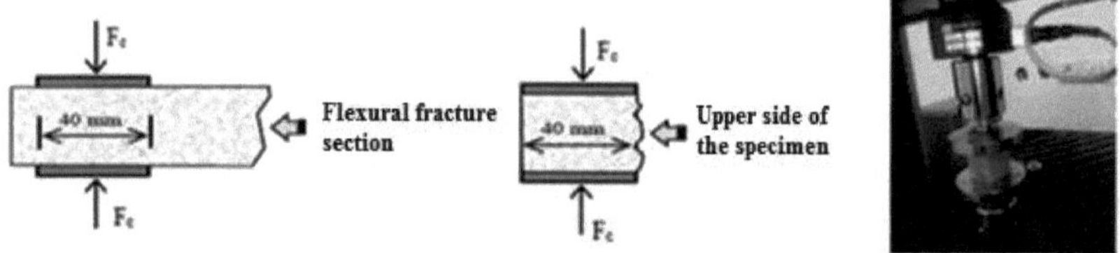

Fig. 25 Compressive strength testing device

A progressive force was applied to the cross-section (4 × 4 cm) of the specimen at the same rate until the breaking load was reached, from which the mechanical compressive strength can be calculated in mega-pascals using the formula below.

Rc = Fc/b*h (**Eq**2)

Then, the samples were also analysed by a Bruker diffractometer of flat technical form at the University of Chouaib Doukkali El Jadida Morocco, using a copper (Cu-Kα) anticathode of wavelength λ equal to 1.5406 A. The samples after grinding were analysed by a HITACHI 2500 C Scanning Electron Microscope.

II.5 Formulation and characterisation of mortar based on hearth ash and magnetised water

II.5.1 Preparation of the mortar

In this study, fire ash (BA) is introduced into the cement mortar formulation by replacing it with different percentages of (BA) with BA/cement mass ratios equal to: 10%. To determine the mechanical properties, the mortars were prepared according to the European standard NF EN 196-1. According to the latter the normal mortar must be composed by mass, of one portion of cement (450 g), three portions of sand (1350 g) and half a portion of water (225 g), i.e. with a water/cement ratio W/C equal to 0.5. The ingredients were mixed according to the requirements of the standard for 4 min in a 5 L mixer at room temperature and relative humidity above 50% (Zalaghi et al.2018). Before starting the mechanical study; the mortars were prepared as prismatic specimens of dimensions 4 * 4 * 16 cm3 as shown in **Figure 26.** The specimens thus formed were stored in the open air. These specimens were then tested after 28 days by flexural and compressive strength. **Table 6** shows the complete formulation of the mortars.

Fig. 26 Fireplace ash mortar samples

Table 6. Formulation of mortar with hearth ash

Elements	Fireplace ash (g)	Cement (g)	Sand (g)	Water (g)
B10	52.07	405	1350	225

II.3.4.2 Analysis of materials to be used

II.3.4.2 .1 Cement

The cement used in the preparation of the fire-ash mortar is the same cement used in the preparation of the fly-ash mortar.

II.3.4.2.2 Sand

The sand used in the preparation of the fire-ash mortar is the same sand used in the preparation of the fly-ash mortar.

II.3.4.2.3 Fireplace ash

The results of the chemical composition analysis of the JLEC bottom ash are given in **Table 7**. It is noted that more than 80% of the chemical constituents of the ash were composed of the constants SiO2, Al2O3 and Fe2O3, with respect to the mineralogical lime

Table 7. Composition of the used fireplace ash

CaO	**SiO2**	**Fe2O3**	**Al2O3**	**K2O**	**Na2O**	**ZnO**	**PbO**	**SO3**	**MgO**	**CaOfree**	**LAW**
1.92	52.07	8.86	23.34	1.9	0.4	0.01	0.01	1.87	1.09	0.29	8.24

The composition of the lower ash contains a maximum of 2% of this element as indicated by the X-ray diffraction technique. This study reveals the existence of two peaks, the first is attributed to quartz (SiO2) and the second corresponds to mullite (**Figure 27**)

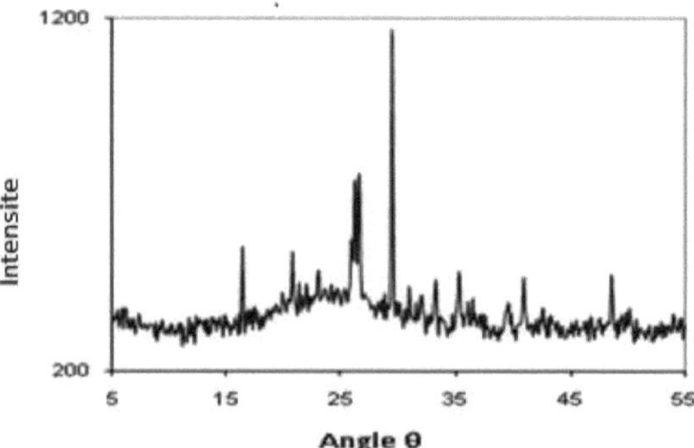

Fig. 27 Mineralogical spectrum of fire ashes

This is due to the mineralogy of the coal used, which generally consists of crystalline silica in the form of quartz and phyllite minerals of the clay (shale) group (Fan et al. 2005). **Figure 28** illustrates the morphological structure of the raw ash obtained before filtration: it appears as spherical particles of irregular size. This result is similar to that reported by (Wei-ling Sun et al), who showed by SEM micrographs that the irregularity observed indicates the presence of porosity. The authors also showed that most of the cement grains were in the form of hollow spheres or spheres filled with smaller spheres. Microcrystals are also observed on the surface of the particles, which may indicate the presence of mullite and quartz. The elements Si, Al, Ca, Na and small amounts of Fe are contained in the plospheres. This is consistent with our mineralogical spectrum shown in Figure 28, the chemical composition of which is shown in Table 4. It can also be noted that the specific surface area measured for this raw hearth ash (400 m2 / kg) remains low compared to that of the fly ash.

Fig. 28 Morphology of hearth ash.

After the mechanical tests, the mortars were also analysed with an XRD diffractometer in the technical platform at the University of Chouaib Doukkali El Jadida Morocco, using a copper anticathode (Cu-Kα) with a wavelength λ equal to 1. 5406 A. The mortars after grinding were analysed by model FTIR infrared transform spectroscopy (Thermo Scientific). We kept the same experimental variables used in the fly ash mortar.

II.4 Formulation of biostimulant based on humic acids

II.4.1 Chemical composition of compost made from manure and coffee grounds

Figure 29 shows the black appearance of the mature compost based on manure and coffee grounds, according to the results obtained from Table 8, this compost is rich in organic matter and nitrogen,

Fig. 29 Compost used for the extraction of organic matter

Table 8: Chemical composition of compost

Chemical element	Content of the element in the compost
Organic matter (OM) content	45%
Nitrogen content (N)	1,5%
C/N	15
pH	7,5
Moisture	< 40%

II.4.2 Extraction of humic and fulvic acids with magnetised water

Soil consists of an organic part that is in the state of humus. Humus is a complex mixture of substances, the most important of which are humic acids, fulvic acids and humin. Humin is insoluble in alkaline reagents and is difficult to evaluate quantitatively, whereas humic acids and fulvic acids can be extracted in a basic medium. Humic acids are insoluble in acidic media, whereas fulvic acids remain soluble in acidic media. This difference in pH properties is used to separate these two categories of substances. The methods used to extract humic substances from compost are generally used to extract organic matter from soil (Parsons et al. 1998). The extraction of humic substances from soil has been the subject of much research. Different techniques have been used. Some authors such as (Wiesemuller et al. 1965, Flaig et al.1970) have made comparisons concerning the extraction of humic and fulvic acids by different solutions. These solutions are soda of 1% concentration used cold and chloridric acid of 2% to 5% concentration. Pyrophosphate and soda solutions can be used in mixture. Extraction and fractionation results vary according to the nature of the solvents and the extraction process (Dabin et al.1976). Other authors have obtained a fairly complete and reproducible extraction of humic matter in soil, using a mixture of sodium pyrophosphate Na4P2O7 (0.1M) (Kononova et al. 1961). For our purposes, we used a solution of KOH with tap water and magnetised water to extract humic and fulvic acids. The choice of KOH is the base generally used to extract organic matter in agrochemistry **Fig. 30 and 31.**

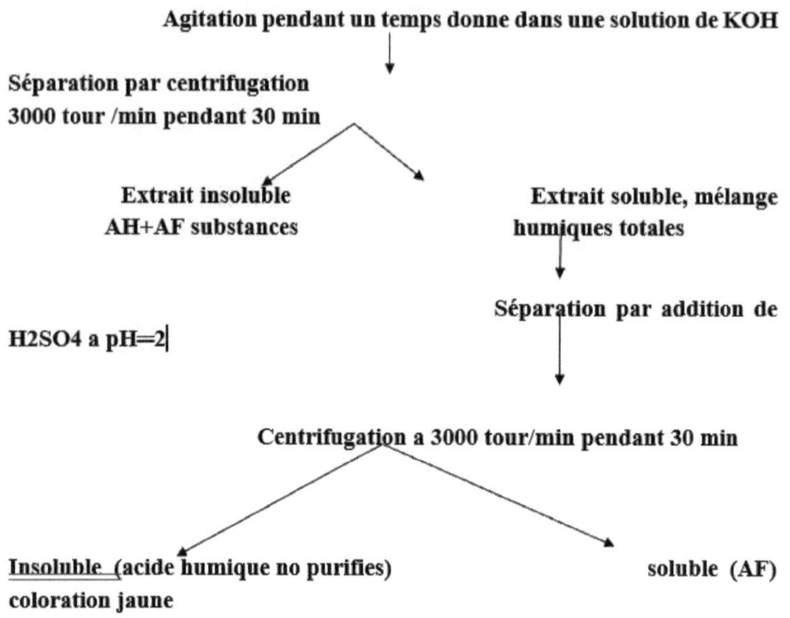

Fig. 30 Protocol for the extraction of humic and fulvic acids from compost based on manure and coffee grounds

Fig. 31 Photo of liquid extract of organic matter from the compost in the laboratory and on a pilot scale

II.4.3 Determination of humic and fulvic acid contents

The methods for the determination of humic and fulvic acids extracted from the soil are numerous and very different in nature. Humic and fulvic acids can often be determined by two different methods, one volumetric and one colorimetric and especially spectrographic, which are faster and more sensitive than the volumetric methods.

II.4.3.1 Volumetric methods

II.4.3.2 Principle

Two methods for the determination of organic carbon by the wet method are used more generally in laboratories: the D'ANNE method and the WALKLEY and BLACK method. Both methods are based on the following principle. A known quantity of an oxidising agent is applied to the soil under well-defined conditions, and the unused excess is then measured. The oxidant used is chromic acid in excess. The determination is carried out cold by means of a ferrous salt solution, which is Mohr's salt. The essential difference between the two methods ANNE and WALKLEY and BLACK lies in the fact that the first involves a hot attack of the soil with boiling for 30 minutes with dichromate in a sulphuric medium, whereas the second takes place cold for 30 minutes of reaction (Walkley et al. 1934). In the cold method, the oxidation of potassium dichromate carbon is incomplete, and the results must be multiplied by a corrective coefficient, as the latter only covers 75% to 77% of organic carbon (Anne et al. 1966). We chose the ANNE method because it is simpler and faster, and above all the dichromate residue is clearer, which makes it easier to visualise the equivalence point during the determination.

II.4.3.3 ANNE method

The carbon in the organic material is oxidised by a mixture of potassium dichromate and sulphuric acid. It is assumed that the oxygen consumed is proportional to the carbon to be determined, according to the equations below. The excess of dichromates is titrated with Mohr's salt (**Figure 32**).

$2K_2Cr_2O_7 + 8H_2SO_4 \rightarrow 2K_2Cr_2(SO_4)_4 + 8H_2O + 3O_2$

$3O_2 + 3C \longrightarrow 3CO_2$

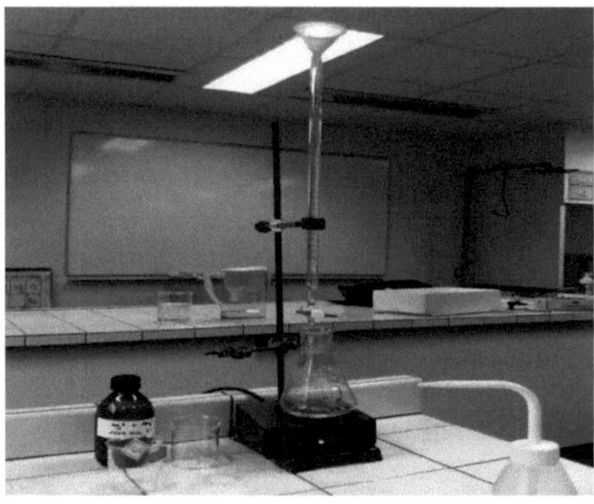

Fig. 32 Determination of liquid extracts by the ANNE method

Generally in the formation of fertilizers based on humic substances, the percentage of humic and fulvic acids is very important. In this study we tried to optimize the method of extraction of humic and fulvic acids by magnetized water in basic medium and also to determine the percentage of organic matter, humic and fulvic acids extracted from compost by using KOH solution of different concentration with magnetized water and tap water for variable times. The determination of organic matter in the compost is done by chromic acid oxidation (8%). The method requires a sufficient excess of dichromate in relation to the carbon in the test sample. As the amount of organic matter extracted from the compost under the new conditions is not known, we proposed to carry out extraction and dosing trials in order to find the best conditions for accuracy. The extraction of organic matter from the compost was carried out with KOH solutions (0.1N, 0.5N and 1N) with tap water and with magnetised water for varying times at room temperature. The amount of organic matter extracted depends on the extraction conditions. With magnetised water and tap water, a solution of KOH (0.1N, 0.5N and 1N). A suitable test sample is 0.25g of dry compost. To extract and determine the humic and fulvic acids under the same conditions, we need a test sample of about 3.27 g of dry compost.

Firstly, we extracted the organic matter by a 0.5N KOH solution with tap water and magnetised water for varying times and at room temperature, then we varied the concentration of (0.1N, 0.5N and 1N) to compare the effect of the concentration on the extraction yield. Then the best conditions were chosen to extract humic acids and fulvic acids.

II.4.4 Determination of organic matter (OM) in compost

II.4.4.1 Procedure: extraction of organic matter

— Weigh 0.39g of dry compost to the crushed area and then pass through a 0.2mm sieve.

— Place the compost in a 250ml Erlenmeyer flask

— add 100ml potassium hydroxide KOH

Place the Erlenmeyer flask on a mechanical stirrer and proceed as follows:

Shake intermittently during different extraction times.

— Put the filtrate in a 100 ml flask

— Complete to 100ml with distilled water

— Take 20 ml from the vial for determination

II.4.4.2 Calculation of percentage organic matter (% OM)

Let V be the volume of mohr solution used for the control assay

Let V1 be the volume of mohr used for the enchantment dosage

Let P be the weight of dry compost (test sample)

Let 0 .196 be the normality of the Mohr solution

The total carbon content of the compost is

Total $\%C = (V - V1) \times 11.7$

We have a dilution of 50 times

Psec= 0.25g

Then: $\%C\ Total = (V - V1) \times 11.7$

To obtain the percentage of organic matter, the total carbon content, obtained by the ANNE method, is multiplied by 1.724. This represents on average 58% of the organic matter: $\%C(OM) = 1.724 \times \%Total$

II.4.5 Extraction and determination of total humic compounds from compost compost

Humic acids and fulvic acids, referred to here as humic compounds, are extracted from the compost using a KOH solution and then separated in an acid medium because of the difference in the behaviour of these categories of substances with respect to pH. The former are insoluble in an acid medium, whereas the latter are soluble.

II.4.5.1 Procedure: extraction of total humic compounds (THC)

- Weigh out 5 g of dry, crushed compost and pass through a 0.2 mm sieve.

Place the compost in a 250ml Erlenmeyer flask

Add 100ml potassium hydroxide KOH

Weigh the Erlenmeyer flask on a mechanical stirrer and proceed as follows:

- Shake intermittently for different times
- Transfer the contents of the Erlenmeyer flask to a centrifuge tube
- Centrifuge for 30 minutes at 3000 rpm
- Filter the extraction solution
- Collect the filtrate in a 1000ml volumetric flask
- Complete to 1000ml with distilled water
- Let V be the equivalent number of ml of the mohr salt solution poured to obtain the turn

Make a control under the same conditions by boiling 10 ml of 8% potassium dichromate for 10 min in the presence of 30 ml of H_2SO_4 concentrate.

- Connecting the tank to a cooling column
- Bring the contents of the flask to a boil for 16 minutes
- Allow the balloon to cool slowly
- Fill up to the gauge line with the flushing water from the flask
- Homogenise the contents of the flask and take 20 ml of solution
- Pour into a 400 ml plain glass beaker
- The 20 ml of solution
- 200 ml distilled water

3-4 drops of phenanthroline

- Place the beaker on a magnetic stirrer with a burette graduated to 1:20 ml containing the 0.2 N
- Titrate by shaking the contents of the beaker from green to red with dark green as an intermediate glue
- Let V be the equivalent number of ml of the mohr salt solution poured to obtain the turn

- Make a control under the same conditions by boiling 10 ml of 8% potassium dichromate for 10 minutes in the presence of 30 ml of H_2SO_4 concentrate

- Make a control under the same conditions by boiling 10 ml of 0.5N potassium dichromate

II.4.5.2 Calculation of the percentage of humic compounds

Let V be the volume of mohr solution used for the determination of the control

Let V1 be the volume of Mohr's solution used for the determination of the sample

Let 0.196 be the normality of the Mohr solution

The carbon content of total humic compounds is expressed as %.

$$C\% = \frac{(V-V1)\times 0.196\times 3\times dilution}{P\times 10}$$

We have a 100 times dilution

Then: $C\% = (V-V1) \times 1.789$

II.4.5.3 Determination of humic and fulvic acids

II.4. 5.3.1 Separation of HA and FA

Pour 80ml of the humic solution into a 200ml Erlenmeyer flask and acidify the solution.

- Centrifuge at 3000 rpm for 30 min

— Remove the supernatant which corresponds to the acid soluble fulvic acids

— Redissolve the humic acids by pouring 50 to 80 ml of Na OH a 0.1 N into the centrifuge tube

— Transfer the humic solution to a 100ml volumetric flask

— Complete to the gauge line

Take 12.5 ml of humic acid solution and pour it into a round-necked flask (12.5 ml)

— Complete up to the gauge line

Take 12.5 ml of humic acid solution and pour it into a round-necked flask (12.5 ml of solution corresponds to 1/10 of the dry weight of the compost test sample)

— Add to the balloon :

— 12.5 ml $K_2Cr_2O_7$

— 40 ml of concentrated sulphuric acid

— Bring the contents of the flask to the boil for

– Then operate as for total humic compounds (AH+AF)

II.. 4. 5.3.2 Calculation of percentage of HA and AF

Let V be the volume of Mohr's solution used for the determination of the control

Let V1 be the volume of Mohr's solution used for the enchantment dosage

Let P be the weight of dry compost (test sample)

This is 0.196 the normality of Mohr's salt solution

The relative carbon content of humic acids is :

$$\%C\ (AH) = \frac{(V-V1)\times 0.196\times 3\times \text{dilution}}{P\times 10}$$

We have a dilution of 50 times

Psec = 3.27 g

So %C (AH) = (V–V1) × 0,89g

The C content corresponding to the fulvic acids is obtained by diffrenece

% C (AF) = C %(CHT) – C% (AH)

Study of the effect of electromagnetic fields on pH

III.1.1 Results and discussion on the pH of magnetised water

Table 5 shows the water analysis values before exposure to the electromagnetic field. Figures 33 and 34 (a, b) show the pH and the variation of the pH as a function of the water in circulation of the water under the influence of the electromagnetic field for 5 min, 10 min and 20 min for two velocities 0.18 (m / s) and 0.6 (m / s). A very remarkable variation of the pH of the water treated by the electromagnetic field (EMF) is noted for the speed v = 0.6 (m / s). This significant variation of the pH of the EMF treated water is due to the Lorentz force which increases with the flow velocity and causes the fragmentation of the hydrogen bonds. Many claims have been made that electromagnetic fields alter the physicochemical properties of water (Li et al.2006, Murad et al 2006). Our experimental results have again verified that electromagnetic treatments modify some of its properties (Guo et al.2006).

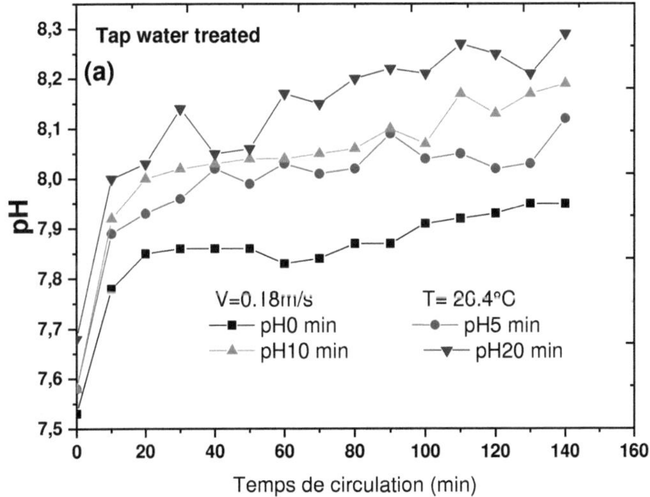

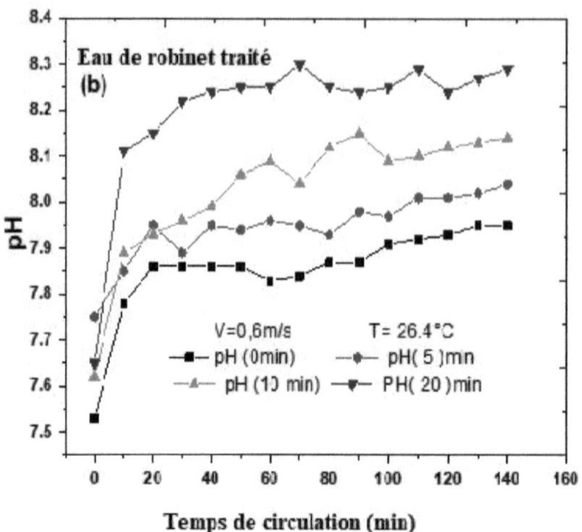

Fig. 33. (a, b) pH changes of tap water exposed to EMF, 5 min, 10 min and 20 min as a function of time, at 26.4°C, flow rate v = 0.18 m / s, v = 0.6 m / s respectively (pH0 untreated).

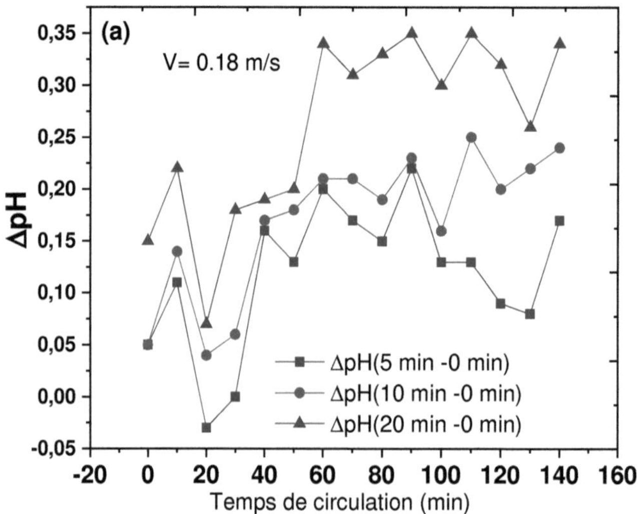

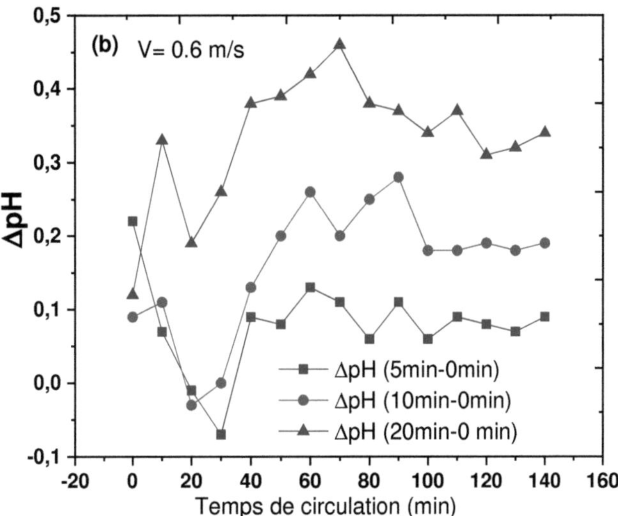

Fig. 34. (a, b) Variation of ΔpH for 5 min, 10 min and 20 min as a function of time, at T = 26.4°C, flow rate v = 0.18 m / s, v = 0.6 m / s respectively (pH0 untreated)).

The calculated pH of the water according to the expression ΔpH = pHEMF-pH0 is carried out at different flow rates, in different time intervals (circulation time) with 5min, 10 min and 20 min being the circulation time. The results obtained show that ΔpH is proportional to the duration of exposure to the electromagnetic field and the flow rate. According to the literature (Wang and al .2012), magnetic field treatments increase the pH of the water and this is consistent with our results. The results of a pH measurement are defined by the amounts of H + ions and OH- ions present in the water. When the quantities of these two ions are equal, the water is considered to be neutral. According to our results, the pH increases slightly as a result of the fragmentation of the water molecule by absorbing the energy generated by the magnetic energy and which helps to break the hydrogen bonds that lead to the formation of H^+ ions and the increase in the number of OH- ions, which results in an increase in the pH of the magnetic water According to our experimental protocol, we have presented the percentage of relative pH during exposure to EMF at T = 26.4°C with a flow rate v = 0.6 m / s at 120 min in figures 35 and 36.

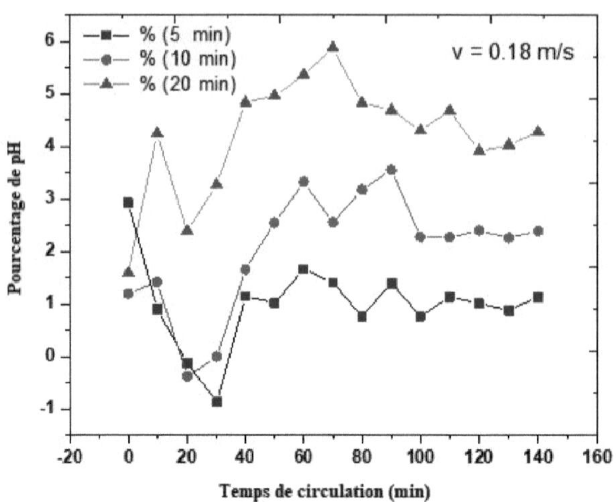

Fig. 35. Percentage change in pH v = 0.18 m / s at T = 26.4°C.

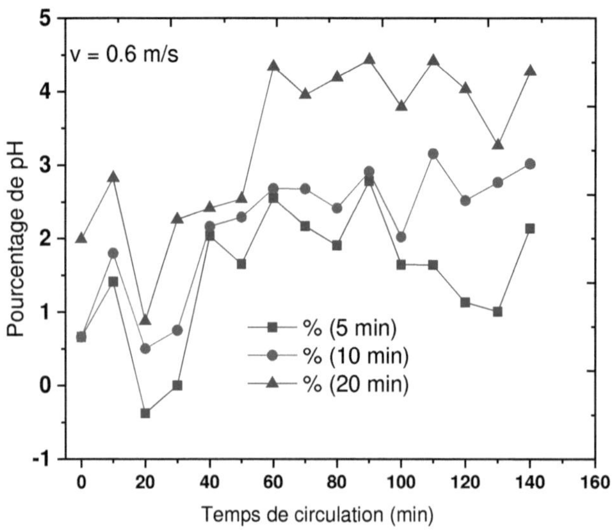

Fig.36. Percentage change in pH at v = 0.6 m / s and T = 26.4°C.

According to these experimental data, the electromagnetic field has a remarkable influence on the physico-chemical properties of the water. The pH is expressed in this regime as follows:

pH=K0×τ (4)

The steady state characterised by an almost constant pH value which shows the evolution of Fig.37. The steady state indicated that the pH reaches a value of 8.25. This parameter is independent of the electromagnetic field. The constant K0 depends on the flow velocity of the water in the pipe where the electromagnetic field is located. The characteristic time τ constant of the electromagnetisation of tap water. It is estimated that the phenomenon of electromagnetisation will be modelled by the following expression using the Origine 8 software following this formula. $pH(t) = A * e^{\frac{-t}{\tau}} + A_0$ (5)

Where: pH0=A0+A

37. Fitting of pH variation of treated and untreated water at 26.4°C.

Data adjustment (v=0 .6m /s at T=26.4°C)

We have also presented the variation of the pH as a function of the time flow. This later representation has two regimes: the transition regime and the steady state regime Fig.38.

In a transient regime, the pH of the water is independent of the effect of the electromagnetic field. The characteristic equation has the following form: pH = K0×τ.

The steady state indicates that the pH reaches a value of 8.25. This parameter is independent of the electromagnetic field. The figure above shows that the variation of the circulation time as a function of the magnetisation time is proportional and this proportionality is a function of the size of the apparatus and the volume of water used in the magnetisation under the effect of the electromagnetic field. These are very important results that have been obtained experimentally with sophisticated measuring instruments. In this study, the use of Aqua-4D devices at different times and flow rates shows that the pH increases slightly with the magnetisation time. It should also be noted that the flow rate and the length of the device (time of passage of the water through the field) have a great influence on the electromagnetic treatment, especially the variation of the pH after the water passes through the aqua-4D device. The variation of the magnetization time as a function of the circulation time is presented in Fig.39. We observe that the changes in the water properties at the time of magnetization depend on several parameters such as the flow velocity and the circulation time under the applied electromagnetic field.Fig. 39 shows the regression of the magnetization time against the real time which allowed us to determine the parameters related to the electromagnetic field generating device. Using this expression, it is possible to determine the length of the device and the volume of water circulating in the closed cycle with the presence of the electromagnetic field. Figure 39 shows the dependence of the pH on the time of circulation of the tap water in the field where it is excited by the electromagnetic field and the evolution as a function of the magnetisation time, i.e. how much magnetisation remains in the water.

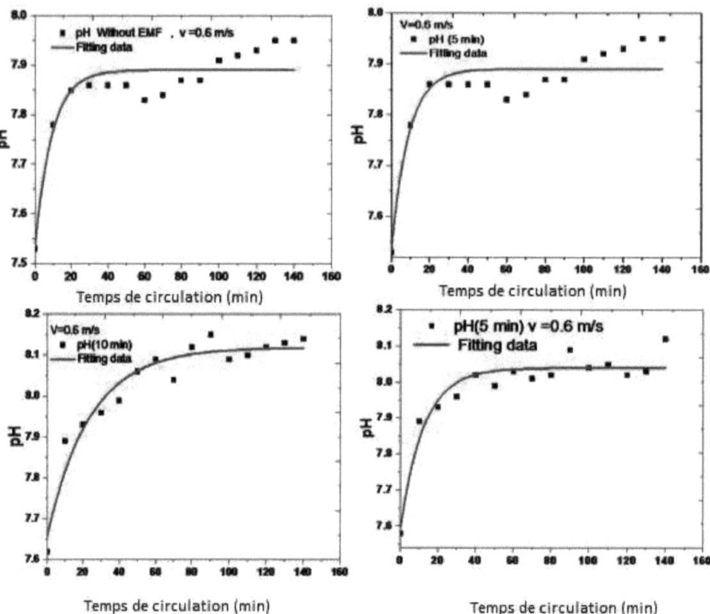

Fig.38. Fitting of pH variation of treated and untreated water at 26.4°C.

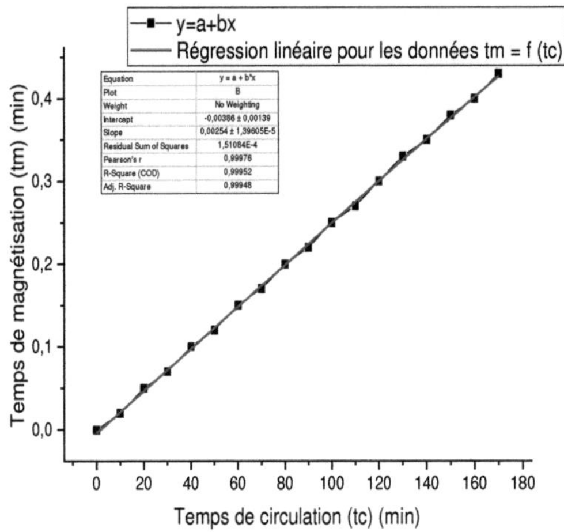

Fig. 39. Variation of the magnetisation time as a function of the circulation time.

IV.1.1 Results of thermal study

IV.1.1.1 heating result

The changes in the evaporated amounts and volume of magnetised and non-magnetised water are shown in Figures 40, 41, 42 and 43. The amount and volume of water evaporated is highest under a flow rate of 0.6 m/s under kinetic conditions. The evaporation rate and volume increase with the time of exposure to the electromagnetic field (EMF).

To highlight these effects, the difference between the amount and volume of magnetised and non-magnetised water evaporated Δm and ΔV at different times is shown in Figures 44 and 45 and Figures 46 and 47. The differences in the evaporated amounts are proportional to the flow rate and the time of exposure to EMFs.The increase in the evaporated amount and volume is higher for magnetised water.

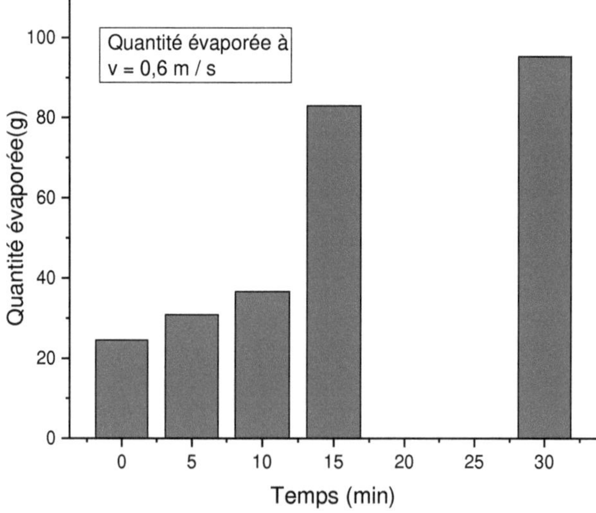

Fig.40. Temperature variation of magnetised water and tap water at v = 0.6 m / s

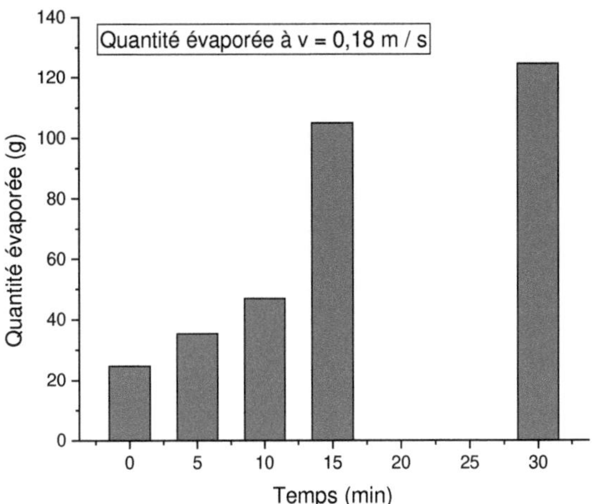

Fig.41 Temperature variation of magnetised water and tap water at v = 0.18 m / s

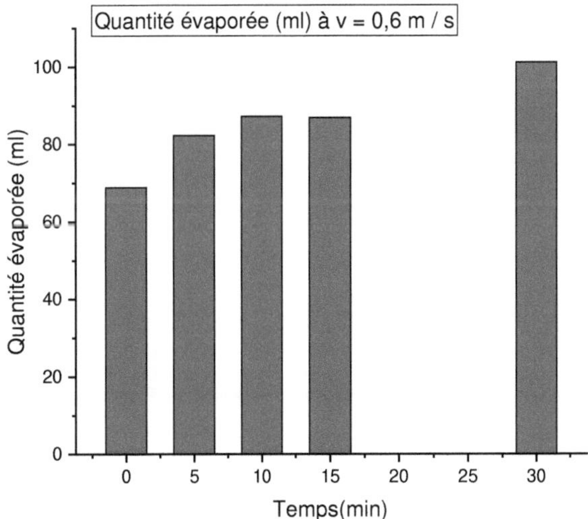

Fig.42.Evaporation of magnetised water and tap water at a flow rate of 0.6 m

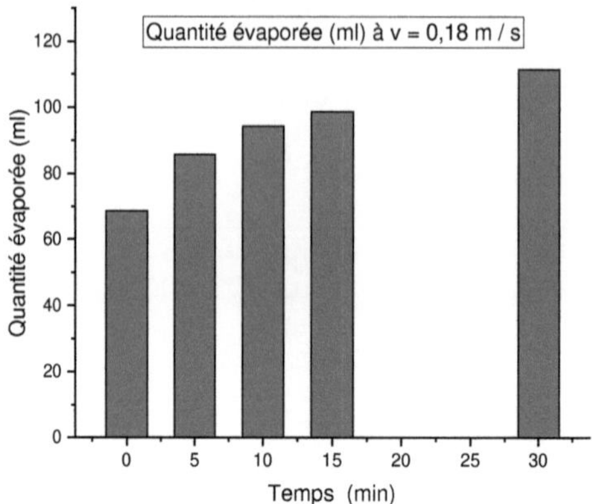

Fig. 43: Evaporation rate of magnetised water and tap water at a flow rate of 0.18 m / s.

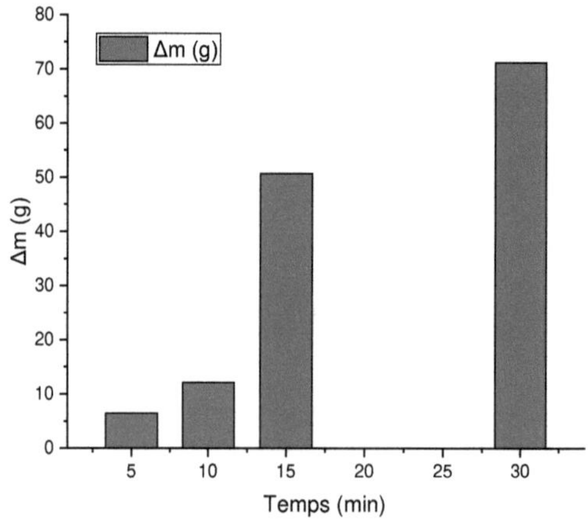

Fig.44.Mass difference of evaporated magnetised water and tap water at v = 0.6 m / s.

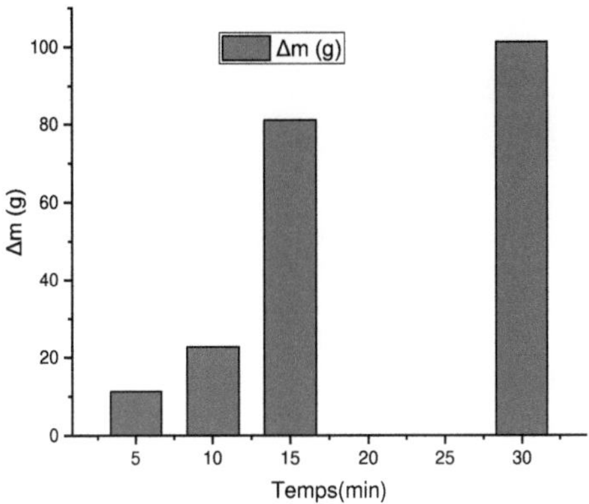

Fig.45 Mass difference of evaporated magnetised water and tap water at v = 0.18 m / s

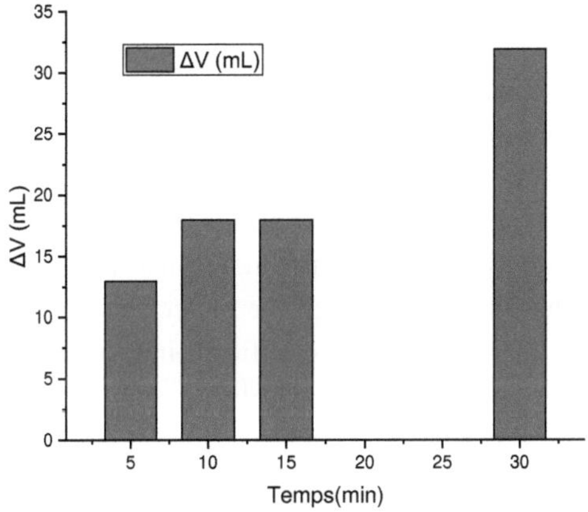

Fig.46.Differences between the volume of evaporated magnetised water and tap water at v = 0.60 m / s.

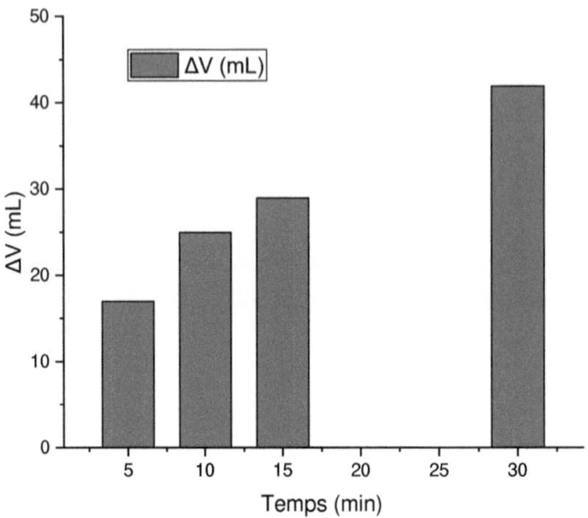

Fig.47 Differences between the volume of evaporated magnetised water and tap water at v = 0.18 m / s.

The heat transfer rate of magnetised water and tap water (control) is presented in Figures 48 and 49. It was found that the heat transfer rate of the magnetised water is higher than that of the tap water. In addition, the process of the effect of magnetisation of water on the evaporation rate is shown in Figures 48 and 49. In the case of non-magnetised water, the water molecules hardly pass into the vapour phase. This implies that the separation of the molecules from each other is very difficult. In this case, there must be a significant interaction force between the molecules, which does not occur as intensely in magnetised water molecules, through the existence of hydrogen bonds. Therefore, it can be suggested that the electromagnetic field has changed the distribution and arrangement of water molecules and their hydrogen bonds. As indicated in previous studies (Toledo et al. 2008; Seyfi et al., 2017),this behaviour may induce a change in structure, as it is known that evaporation is a process in which water molecules escape from the liquid to the gaseous state.

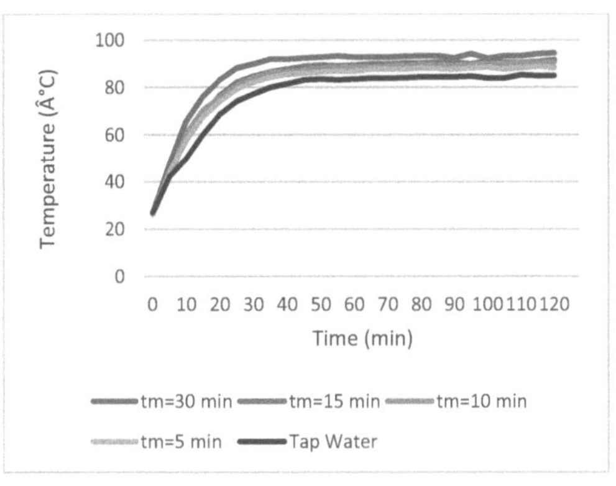

Fig.48.Temperature variation of magnetised water and tap water at v = 0.6 m / s.

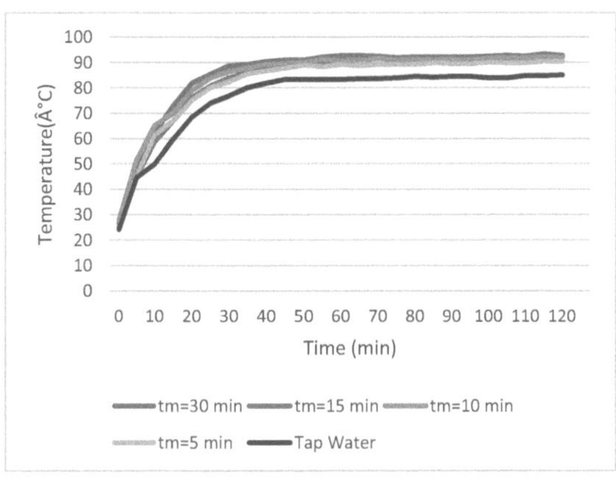

Fig.49.Temperature variation of magnetised water and tap water at v = 0.18 m / s.

Figures 50 and 51 show that there is a difference between the temperature of tap water and magnetised water as a function of the water flow rate and the effect of the electromagnetic field at a velocity of 0.6 m /s and 0.18 m/s. The destruction of hydrogen bonds is the cause of the Lorentz force of the moving water pipes and the interface (Seyfi et al., 2017).

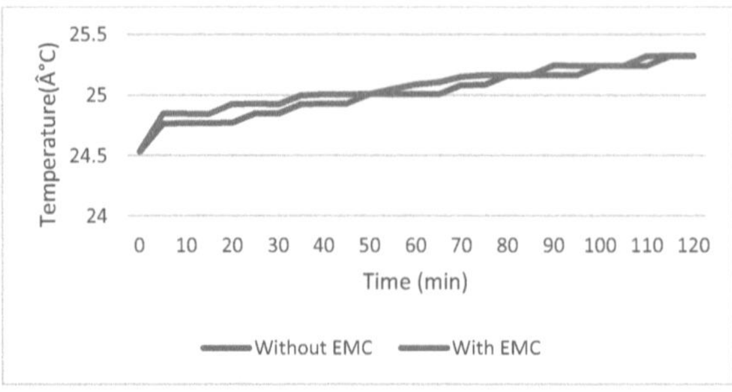

Fig.50. Evolution of the temperature of tap water under the effect of the electromagnetic field in kinetic conditions at v = 0.18m / s.

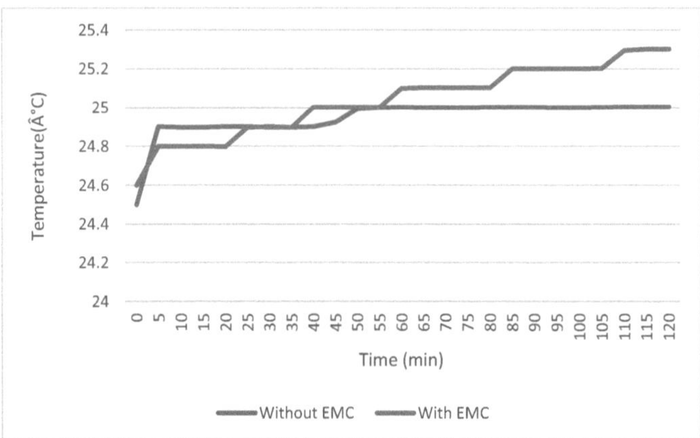

Fig.51. Evolution of the temperature of tap water under the effect of the electromagnetic field in kinetic conditions at v = 0.6m / s.

IV.1.1.2 The effect of the electromagnetic field on water cooling

The cooling tests were carried out to investigate the effect of the electromagnetic field on the cooling rate of the water. After heating the water samples, continuous temperature measurements were made to assess the cooling rate of tap water (NMW) and magnetised water (MW).

Examination of Figures 52 and 53 shows that the cooling rate of the (NMW) is faster than that of the magnetised water generated by the electromagnetic device process. This experiment shows that the technology of the electromagnetic devices influences the percentage of heat transfer during the cooling phase. The difference between tap water and magnetised water results from the disruption of hydrogen bonding and ionic movement due to the electromagnetic field generated by the electromagnetic device (Szcześ et al.2011).

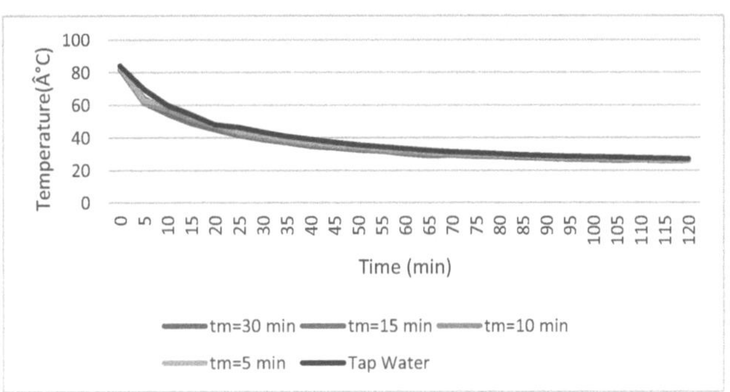

Fig.52.Temperature variation during cooling of magnetised water and tap water at v = 0.18 m / s.

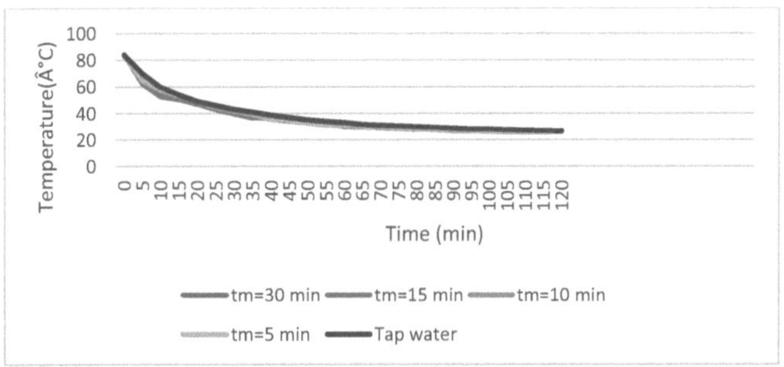

Fig.53.Temperature variation during cooling of magnetised water and tap water at v = 0.60 m / s.

IV.1.2 Calculation of the heat capacity by mass (Cp)

Table 1 shows the results for the heating temperatures for a given water body.

Table 1. Variation of the temperature difference when cooling distilled water.

Water body	θ0	θ1	θf	CP (distilled water)
m1 =130g	18.7	50°C	36.7°C	15.29jk-1g-1
m2 =130g	19.3	50°C	36.5°C	15jk-1g-1
m2 =130g	18.4	50°C	36.3°C	15.98jk-1g-1

Table 2. Variation in temperature difference when cooling tap water.

mass	θ0	θ1			CP (Tap water)
m1=130g	18.7	50°C		33.6°C	4.92 jk-1g-1
m2=130g	19.3	50°C		33.9°C	4.82 jk-1g-1
m3=130g	18.4	50°C		33.3°C	4.1 jk-1g-1

Table 3. Calculation of the heat capacity Cp of magnetised water

mass	θ0	θ1		θf	CP (magnetized water)
m1 =130g	18.7	50°C		33.7°C	3.99jK-1g-1
m2=130g	19.3	50°C		33.5°C	3.06jk-1g-1
m3=130g	18.4	50°C		34°C	4.46jk-1g-1

The calculation of the heat capacity of the (MW) of m0 = 130 g can be done by considering distilled water as a reference using the following equation:

$$C_{MW} = \frac{\mu dis \times Cdis \times (\theta_0 - \theta_f)}{m_1(\theta_f - \theta_1) + m_0(\theta_f - \theta_0)} \quad \textbf{(Eq.5)}$$

With:

$$\mu = \frac{m_1 \theta_1 - \theta_f}{m_0(\theta_f - \theta_0)} \textbf{(Eq.6)}$$

IV.2.2 Discussion of heat capacity

The heat capacity values of tap water and magnetised water are 4.6133 jk-1 g-1 and 3.8366 j-1 k g-1 respectively (Cmoy (magnetised water) < Cmoy (tap water)). From this result and Fig.54, it can be seen that the heat capacity of water decreases with a percentage of 20.24% under the action of the electromagnetic field. The exposure of water to the electromagnetic field increases the heating rate and decreases the cooling time, compared to tap water. These results are consistent with those reported in the literature (Alimi et al., 2009; Nakagawa et al., 1999; Seyfi et al., 2017; Szcześ et al., 2011).

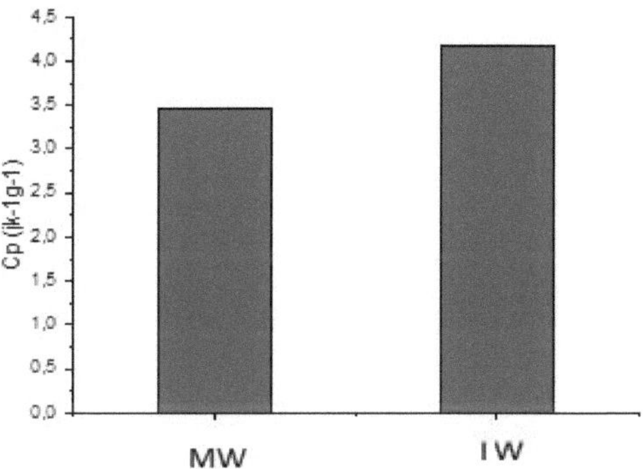

Fig.54. Heat capacity of magnetised water and tap water

IV.1.3 Surface tension calculation

Table 9 and Figure 55 show the masses of the tap water drop and the magnetised water and their surface tension at different temperatures.

Table 9. Surface tension values of tap water and magnetised water at 26.5°C	
Mass of tap water drop (g)	0.05268
Surface tension of magnetised water (N / m) (T = 26.5 ° C)	0.082290
Mass of magnetised water drop (g)	0.041688

Surface tension of magnetised water (N/m) (T = 26.5 ° C) 0.065120

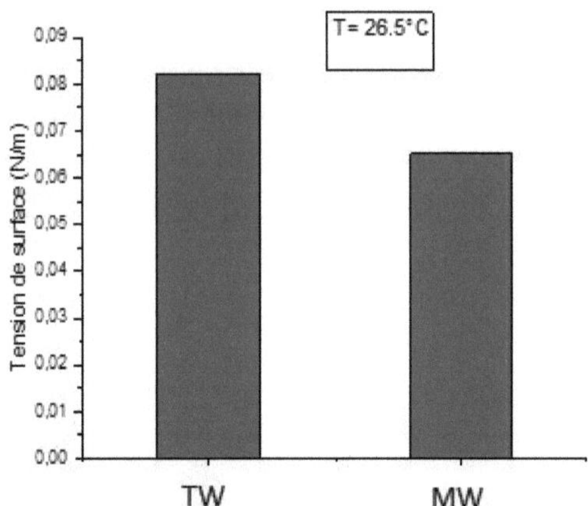

Fig.55 Surface tension of tap water and magnetised water

IV.2.3 Surface tension discussion

The results of the different surface tension measurements of tap water and magnetic water recorded in Table 9, show that this physico-chemical quantity decreases when tap water is exposed to electromagnetic fields produced by electromagnetic device technology. Comparing the evolution of the two types of water, it can be seen that the magnetic field significantly reduces the surface tension. The surface tension of tap water flowing at a flow rate of 0.2 m / s and exposed for 15 minutes to the electromagnetic field at 26.5 ° C decreased by a percentage of the order of 26.36%. the results obtained are in agreement with those of (Amor et al.2017), who reported that the magnetic field can decrease the surface tension of water.

IV.2.4 Correlation between surface tension and heating and cooling time

The surface tension of the (TW) flowing at a speed of 0.2 m/s for an exposure time of 15 min to the electromagnetic field at 26.5°C is decreased by a percentage of 26.36%. The heat capacity of the tap water exposed to the electromagnetic field produced by the electromagnetic device is decreased by 20.24%.

These results are consistent with those of (Seyfi et al. 2017, Amor et al. 2017, and Szcześ et al.2011) where it was shown that under the effect of the magnetic field, the surface tension decreases and the proportion of water evaporation increases. This effect can be interpreted in terms of weakening, breaking of hydrogen bonds and disruption of the gas/liquid interface, thus facilitating the release and escape of water molecules to the vapour phase.Other references (Seyfi et al. 2017) also suggested that, following magnetic field treatment, the structure of water was altered, i.e. more monomeric water molecules or weakened water bundles were present for a period of up to 40 minutes, which might be related to the memory effect.

ELECTRICAL STUDY

Electrochemical impedance spectroscopy (EIS) is used in this study to determine the magnetisation of the electromagnetic field of tap water. In previous studies, there was no identification of the region where the effect of the electromagnetic field is manifested on tap water. This section presents the analysis and study of the electrical parameters, obtained after fitting the experimental data of impedance spectra by an electrical circuit. The impedance of the studied system is given by:

$$Z_C(?)? \quad \frac{1}{j?C}, \quad \text{j est le nombre imaginaire}, j^2=-1$$

(Eq.7)

$$??(??) = \frac{1}{??(??)^{??}} = \frac{1}{??} - ??^{-??}\left(\cos\frac{-???}{2} + ?\sin\left(\frac{-p\pi}{2}\right)\right)$$

(Eq.8)

The impedance of an EPC is a function of the angular frequency ω / rad s-1, the adjustment factor p, the imaginary number j and the amplitude T / sp-1. For n = 0, it is equivalent to a resistor (Q-1 = R) and for n = 1, the EPC is equivalent to a normal capacitor (T = C). The phase angle of this EPC element over the entire frequency range is equal to - nπ / 2.

V.1 Impedance analysis of magnetised water and tap water

In the electrical circuit, the polarisation of the electrodes is not directly similar, but can be schematised as a combination of some components (Ueno et al . 2012): a constant phase element (CPE) (Mghaiouini et al., 2020b) with Warburg impedance (W) in parallel to consider the displacement of the ions The R and W impedances show the total characteristics of the probe distribution in the electrolyte. These parameters are not able to follow the field distribution as closely and do not show the same polarisation rate (Tijing et al., 2010). The build-up of water molecule clusters can be detected by EIS in three ways: The growth of Raq is explained by the smaller number of free ions, the decrease in electrode polarisation for the same purpose and the impotence of the larger water cluster to help the electric field and form films, to appear as a change in the CPE and W characteristics, respectively. (Fig.56 (b)) shows the theoretical curve (line) and the experimental curve of tap water (dots). The contribution of electrode polarisation is confirmed by modelling the curves using an equivalent electrical circuit (Fig.57) without Warburg and PEC impedance. These simulations are shown as dashed curves in (Fig.56a & b). The contributions of electrode polarisation are distinct in these figures.

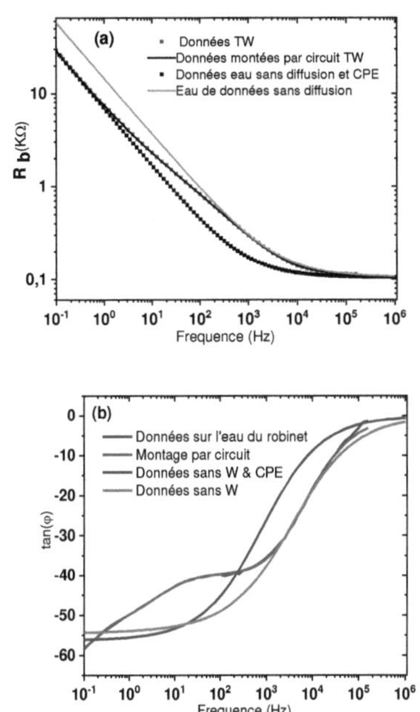

Fig. 56: Variation of volume resistance (a) and phase angle (b) as a function of frequency

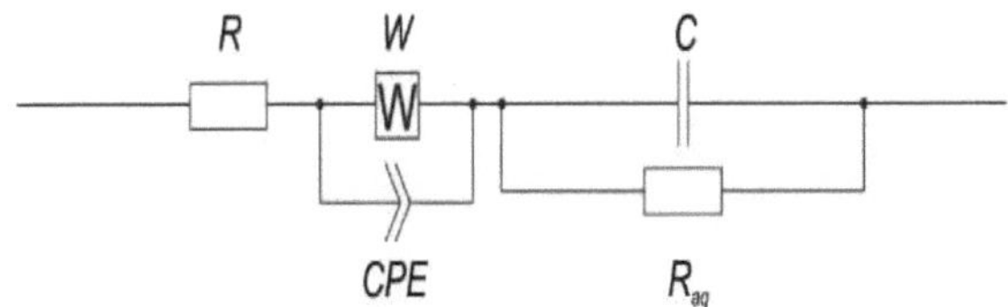

Fig.57.Modeling of equivalent electrical circuits of EIS data.

V.1.1 Study of permittivity, complex conductivity and phase angle

The evolution of the imaginary part (ε ' ') of the complex permittivity ε * (ω) as a function of frequency at different samples is shown in Fig.58 (b), including all the samples that showed similar behaviour. The imaginary part decreases with increasing frequency at low frequencies. At high frequency (HF), this plot shows a relaxation peak between 102 and 105.

The high frequency spectrum states that the relaxation peak between the values of 102 and 105 at low frequency seems to be related to the scattering mechanism. The relaxation mechanism is related to the semicircle observed at HF, and the diffusion mechanism is related to the line at LF. In general, the scattering mechanism is related to the low frequency (LF) line. The intersection of the x-axis and the diameter of the semicircle represent the dielectric strength (Δε). The origin of this relaxation process seems to be at Hz in the imaginary part of the conductivity. The relaxation frequencies (τfr) shift to high frequencies below 5 minutes of temporal magnetization in the flow rate of 0.6 m / s. Above this amount, all relaxation frequencies are almost constant for all studied samples. At low frequency (LF), no clear relaxation peaks were identified in the imaginary part of the complex permittivity. The Nyquist diagram shown in Figure 58 (c) reveals the change of the imaginary function (ε ' ') as a function of the real function (ε '). All compositions showed similar behaviours, indicating that there are two complicated processes in the evolution of dielectric spectra. The first process at HF represents a relaxation behaviour of the electrode polarisation, while the second is attributed to the condensation of positively charged particles in the presence of mineral ions on the surface part of the electrode (negative charge), when the numerous aggregations of the cluster water occur. The relaxation is shown in the inset of Figure 58 (c) representing the ε "as a function of ε".

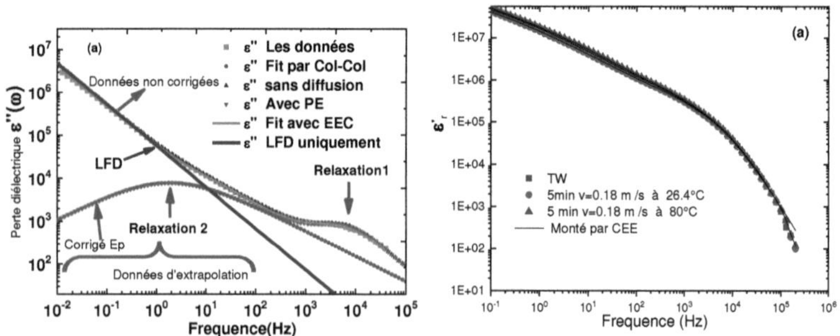

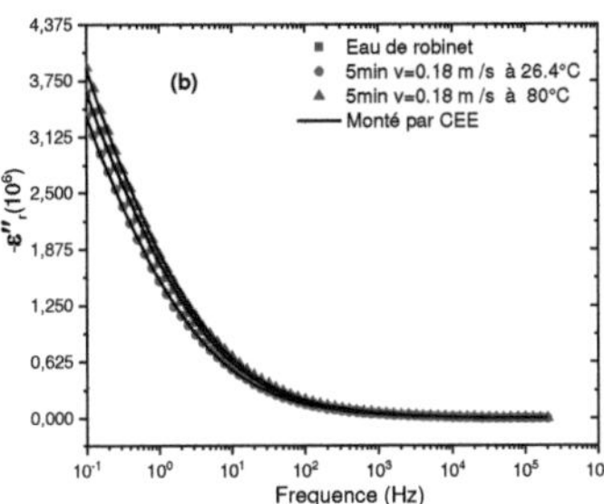

Fig.58. Real (a) and imaginary (b) parts of the permittivity complex as a function of frequency and (c) εr "vsεr". The magnetisation time is 5 min, v = 0.18 m / s at a temperature of 26.4°C and 80°C.

The impedance spectra (Fig.58) show the real and imaginary function of the complex conductivity, and the imaginary function, according to the real part to analyse the change of conductivity with frequency. In addition, the analysis of the imaginary part as a function of frequency was carried out in order to trigger relaxation frequencies proportional to the relaxation times, i.e. the time required for the material to reach an equilibrium state

The variations in total conductivity shown in Figure 59 (a) are similar. The dominant influence is that of temperature. The scale variations shown in figure 59 (b) are also similar. The evolution of the real conductivity as a function of the imaginary conductivity, shown in Figure 59, shows a similar behaviour. This representation has been used to visualise the resistance variation that depends on magnetisation, using the electromagnetic field produced by electromagnetic device technology. The variation of the volume resistance shown in Fig. 59 (d) is quite identical. The curves in Fig. 59 show the magnetization-dependent phase angle variation using the electromagnetic field produced by the technology.

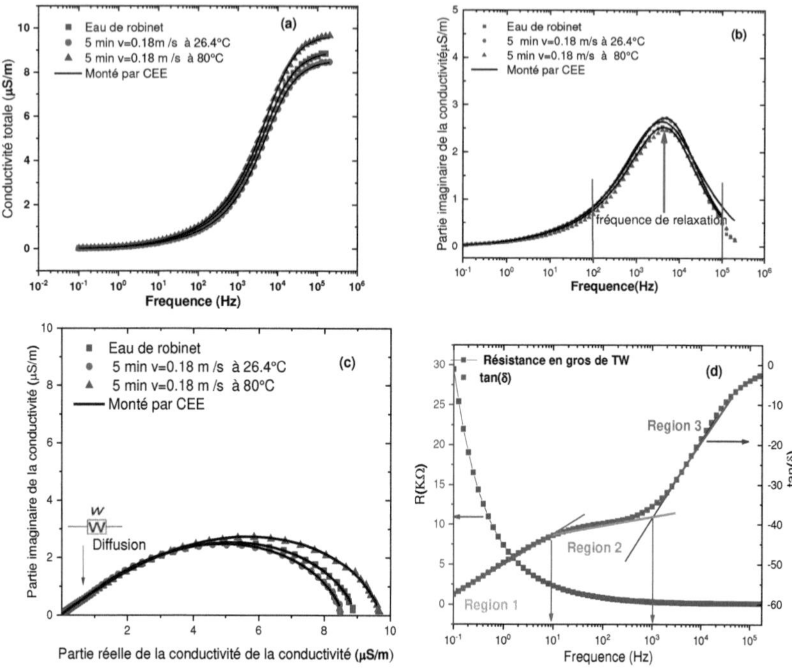

Fig. 59.Variation: of the total conductivity and the imaginary part of the conductivity is appropriate to the frequency, of the imaginary part versus the real part of the complex conductivity under the conditions of 5min, v = 0.18m / s , T = 26.4 ° C and 80 ° C

Dielectric spectroscopy measurements have shown that the phase angle appears to be dependent on the amount of electromagnetic field applied. The loss factor increases with increasing moisture content. Higher magnetisation results in polarisation and higher mobility.

This allows molecules, dipoles and ions to assist the electric field and thus absorb energy. A high loss factor and dissipative processes in conductivity are attributed to the movements of ions and cations and the magnetised cluster (Dissado and Hill, 1989). Similar results were found in a previous study (Kaden et al., 2013). Energy dissipation, resulting from the energy absorbed during exposure in the electromagnetic field by the polarisation of the cluster water density, is the physical origin of the PBS in this study **Fig. 60**

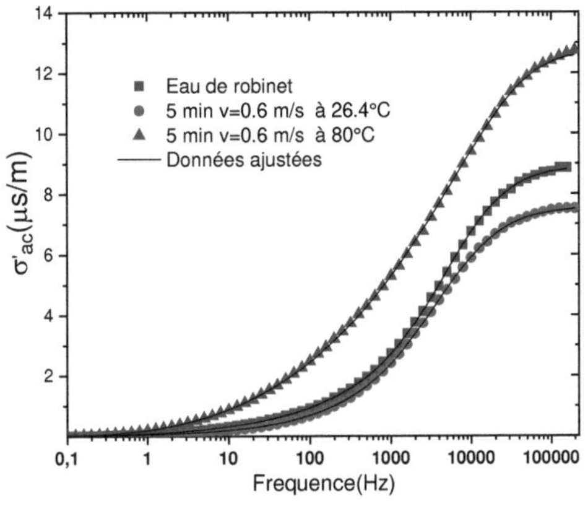

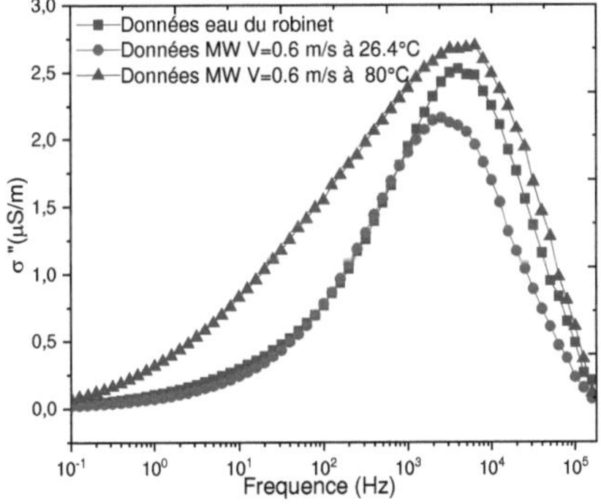

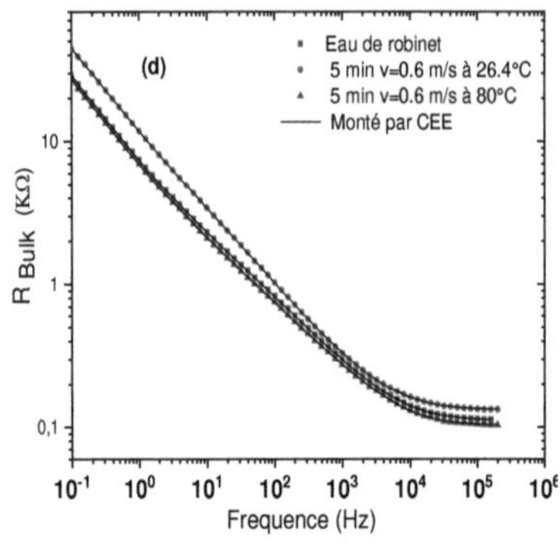

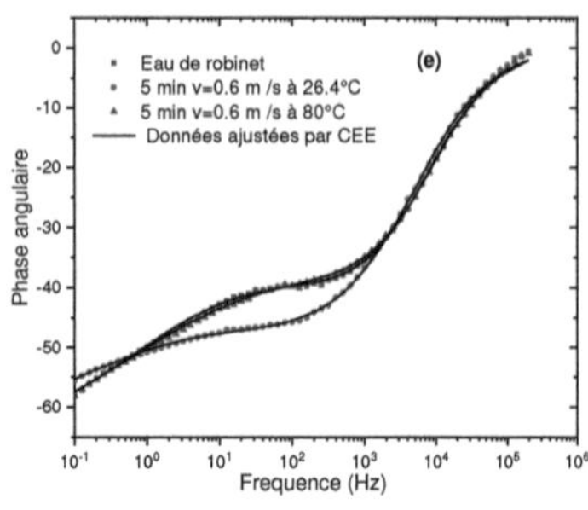

Fig.60.The real (a) and imaginary (b) part of the conductivity complex as a function of frequency and resistance (c) of the bulk phase and angle as a function of frequency for the magnetization time is 5 minutes, v = 0.6 m / s at a temperature of 26.4 ° C & 80 ° C.

The variations in total conductivity shown in Figure 61 (a) are identical. The dominant influence is caused by temperature. The characteristics of the total conductivity curves shown in Figure 61 (b) are also similar. Curves have been used to visualise relaxations independent of magnetisation. The variations in total conductivity shown in Figure 61 (c) are also similar. The variation that depends on the magnetisation induced by the electromagnetic field produced by the Aqua electromagnetic device technology has been visualised. The evolution of the total conductivity asserted in figure 61 (d) is similar. Figures 61 (e) and 61 (d) show the phase angle used. The variation of the phase angles is divided into three regions depending on the frequencies. The charge is higher at high frequencies > 1000 Hz, and the capacitive nature is noticeable at these frequencies. This result correlates with the electrochemical mechanisms of creating the electrical double layer at the interface (Chahid et al. 2013; Laurati et al., 2012). Furthermore, it implies that a large amount of [water/anions/cations]/is present in the metal surface.

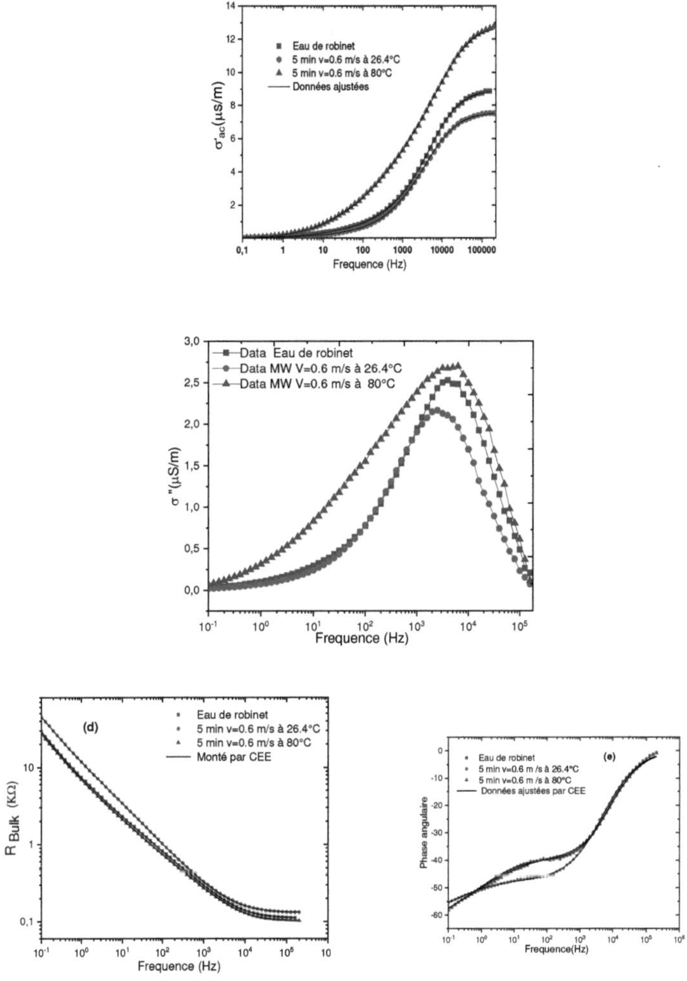

Fig.61.Variation of conductivity with frequency (a), Variation of the imaginary part of the conductivity

as a function of frequency, real vs. imaginary part of the complex conductivity (c) & evolution of the phase angle as a proportion of frequency for tap water 5 min of temporal magnetisation at 26.4°C, 80°C with v = 0.6 m/s.

V.1.2 Impedance and diffusion of the magnetised water molecule / tap water

In order to evaluate the impedance data, an electrical element analysis was discussed in the previous section. Hereafter, methods for calculating the diffusion coefficient and the water volume fraction are considered. The diffusion coefficient

characterises the amount of damaged water, which can also be measured by the EIS technique. The diffusion coefficient can be measured by applying the Fickian method. This method confirms that the constant is not proportional to the concentration. However, for many cationic and anionic solvents present in treated water, the diffusion coefficient is proportional to concentration (Bonora et al. 1996). Although in many
In this case, no experimental evidence exists, it is assumed that the electromagnetic absorption process is Fickian.

A second approach was used to confirm that the variation in water content is proportional to the variation in capacity. It is assumed that all types of water behave in the same way with respect to effects such as polarisation and water/water distribution interactions. For a Fickian process, the relationship between the amount of absorption energy and the diffusion coefficient can be found in the textbooks (Reuvers et al., 2012) Fig.62. The Fickian absorption curve of the theoretical weight is equal to the measured capacity. To calculate the diffusion coefficient, the ratio of the change in capacity was measured at the beginning of the experiment (Weisenberger et al.1989) . The water volume ratio is often evaluated by the Brasher-Kingsbury equation in the electrochemical impedance domain (Frank et al.). In this mathematical relationship, the electromagnetic waves, influencing the water, are calculated from the water cluster capacitance Cf (with MFE) / F at any time and the capacitance upstream of the experiment Cf (without MFE) / F. Taking the dielectric constant of water to be 80, the Brasher-Kingsbury equation is given by:

$$\frac{\log\left(Cf\ (\text{with EMF})/\ Cf\ (\text{withoutEMF})\right)}{\log(80)} \qquad \text{(Eq. 9)}$$

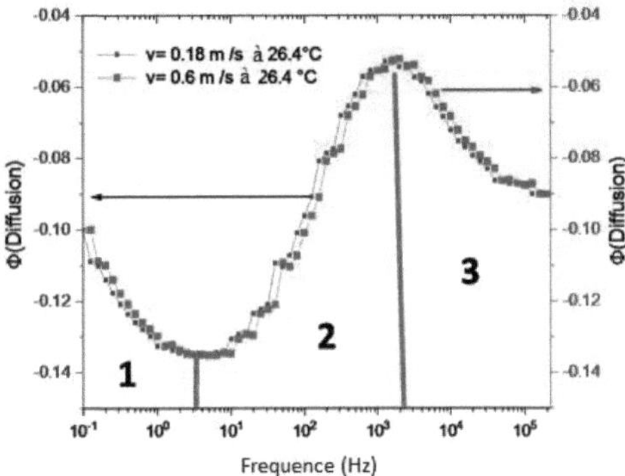

Fig.62.Variation of the diffusion coefficient using the Brasher - Kingsbury equation

V.2 pH test results by dielectric spectroscopy and impudence

V.3 Impedance spectroscopy studies

Impedance spectroscopy has been widely used to examine the properties of electrical materials and electrochemical systems. Typically, impedance spectroscopy measurements are performed to identify physical processes and evaluate various electrical parameters appropriate for the electrical system under study.

The AC conductivity was estimated from the real and imaginary parts of the spectrographic impedance over a magnetisation time of the study range using the following equation30:

$$\sigma'_{ac}(\omega) = \frac{e}{S} * \frac{Z'}{Z'^2 + Z''^2} \quad (1)$$

The frequency dependence of the AC conductivity in tap water at different times of magnetisation is shown in Figure 5.

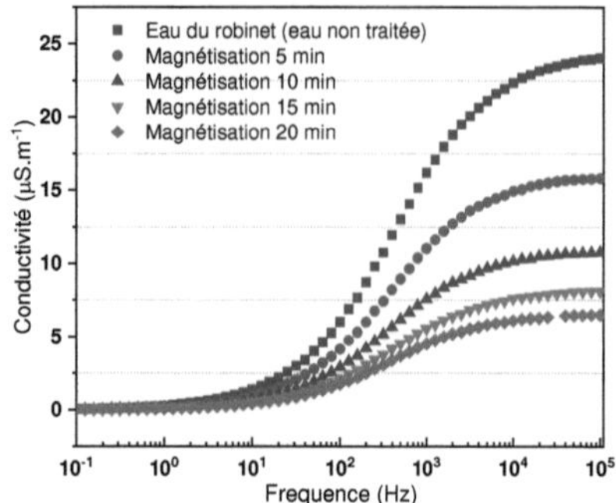

Figure 63: Variation in conductivity of all samples

The analysis in Figure **63** indicates that conductivity is independent of frequency in the low frequency regime. Indeed, magnetising the water can increase the mobility of the water and thus its conductivity, since the resistance decreases with increasing frequency.

In addition, the conductivity of magnetised water depends on the magnetisation as shown in Figure 64. Note that the frequency is between 100 Hz and 100 KHz. The increase in frequency is associated with an increase in conductivity. It is worth mentioning that the electromagnetic field has clear and tangible effects (Figure 64). Furthermore, it can be suggested that the electromagnetic field has a direct effect on the dipole moment of the water molecule: thus, on water clusters, the bonds of the compositions become closer to tap water and the hydrogen bonds become stronger.

Figure 64. Variation of the imaginary part of the conductivity as a function of frequency

Figures 64 and 65 show the real part of the conductivity and the imaginary part as a function of the real part of the conductivity. These figures show that the relaxation frequency of magnetised water is unchanged, implying that there is no deformation in the structure of the water when it is under the flow of an electromagnetic field. It can therefore be suggested that this behaviour is the result of the rearrangement of the water clusters.

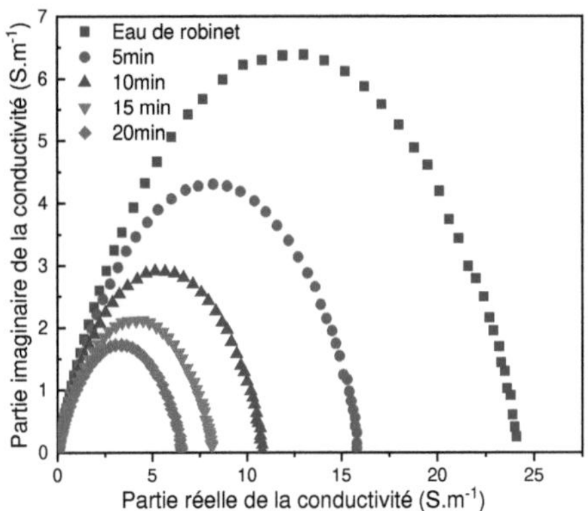

Figure 65. Evolution of the imaginary part as a function of the real part of the conductivity.

V.3.1 Discussion of results and equivalent electronic circuits

Figure 66 shows the conductivity cell with a polarisation effect and its electrical model and the resistance of the liquid column. The Rc is attributed to the resistance of the liquid column (Ueno et al.2012), the RP1 represents the polarisation resistance of the electrodeposition and the RP2 explains the polarisation resistance of the negative electrode The CPE is more versatile in approximating dispersive impedance spectra than the "ideal" capacitance C (Elmelouky et al.2015). Below is an equation that defines the impedance of an EPC (Eq. (2)):

$$Z_{CPE} = Q^{-1}(j\omega)^{-n} \quad (2)$$

Where Q is a real constant independent of the frequency, n is an exponent that depends on the frequency. For n values, the EPC is replaced by a classical element of electrical circuits: capacitance C (n = 1), resistance R (n = 0) and inductance L (n = -1). The value of n = 0.5 is interpreted as the Warburg impedance (W). The other values of n approximate some other types of frequency distribution behaviour of C, R, L or W with the distributed parameters. The capacitive semicircular dispersion is thus related to heterogeneity due to surface roughness, impurities or dislocations (Murad et al. 2006 , Cai et al .2006 , Seyfi et al.2017) fractal structure(Szkatula et al .2002 , Zielinski et al.2017, Liu et al.2016) inhibitor adsorption and porous film formation (Quinn et al.1997, Seyfi et al.2017) . The fitting and adjustment of the impedance spectra was carried out using Zview version 2.2 software, in coordination with the equivalent circuit shown in Figure 66.

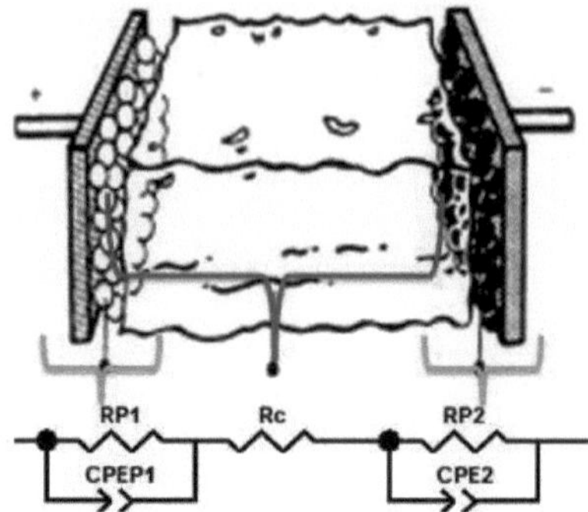

Figure 66. Electrical equivalent circuit used to adjust impedance data.

The percentage of the electromagnetisation coefficient (Magnetisation () + Polarisation () can be estimated by the following equation (Ueno et al.2012):

$$\left(\frac{\Delta R}{R}\right) * 100 = \left(\frac{R-R0}{R0}\right) * 100 \qquad (3)$$

Where R and R0 are defined as the charge transfer resistance without and with treatment, respectively. Figure 67 shows the variation of the electromagnetisation coefficient with frequency for different treatment times. Figure 68 provides a comparative analysis of the conductivity of magnetised water and tap water.

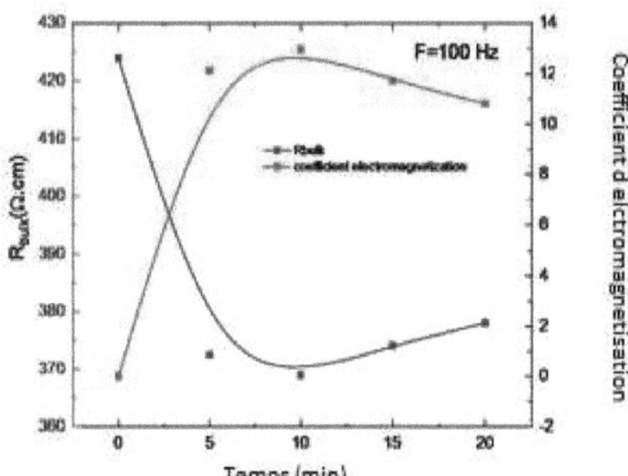

Figure 67. Variations in the electromagnetic field effect of water with increasing exposure time and resistance of the liquid column.

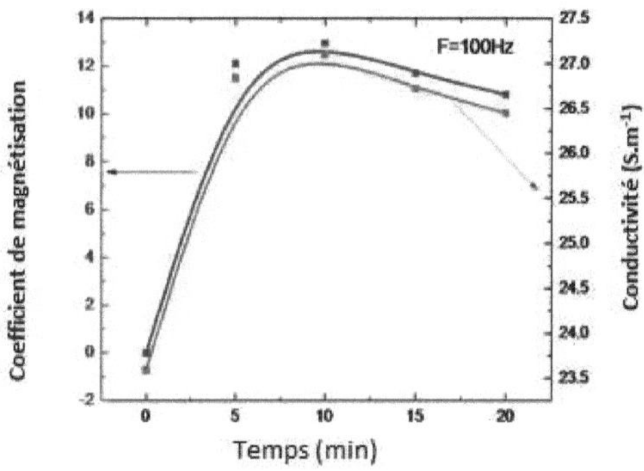

Figure 68. Variations in the effect of the electromagnetic field on the water with time and evolution of the conductivity of the liquid column

OPTICAL PROPERTIES

VI.1 Discussion and results of some optical properties of magnetised water

It is known that the infrared spectrum of water gives an idea of its atomic and molecular structure. Therefore, tap water was exposed to a weak electromagnetic field (40 nT) for 20 minutes of circulation at two speeds: 0.18 m s-1 and 0.6 m s-1. The effect of the electromagnetic field on the optical properties of the magnetised water compared to distilled water and tap water as controls was analysed by the FTIR spectroscopy technique as shown in Figures 69, 70, 71 and 72. After examination of the recorded infrared spectra, it was found that the electromagnetic field influences the intensity and width of the OH band and that its peak shifts slightly towards lower wave numbers. Both observations can be explained by the fact that the magnetic field modifies the structure of water, i.e. the structure of its hydrogen bonds. For example, in the literature (Chang et al.2006) molecular dynamics simulations have revealed that a field ranging from 0 to 10T causes a slight increase in the amount of hydrogenated bonds (0.34%). This results in larger water aggregates and a more stable water structure.

Based on theoretical calculations (Monte Carlo simulation),(Toledo, et al.2008) suggested that the magnetic field breaks the intra-group hydrogen bonding and strengthens them between groups. Water, being dipolar, can be partially aligned by an electric field and this can be easily shown by the movement of a water jet by an electrostatic source (Bramwell et al.2016). (Szczes et al.2011) stated that the magnetic field increases the number of hydrogen bonds, which leads to an improvement in the arrangement and stability of water molecule colloids and generates molecules with hydrophobic behaviour. (Xueyun et al.2016) In addition, the magnetic field in static mode influences the optical properties of water [26]. The OH stretching band around 3400 cm-1 is often decomposed into a few Gaussian peaks that supposedly correspond to water molecules bound together by different hydrogen bonds. In liquid water, the molecular stretching vibrations shift to higher frequencies (as the hydrogen bond weakens, the covalent OH bonds strengthen, making them vibrate at higher frequencies), while the intermolecular vibrations shift to lower frequencies and become more solid and rigid (Choe et al.2016). Hydrogen bonds between water aggregates are accentuated, favouring stretching vibrations and minimising bending. The dependence of this band on hydrogen bonding makes it sensitive to changes in the structure of liquids and solutions under the effect of the electromagnetic field, although this importance is apparently often overlooked. This band is useful for indicating the relative strength of the hydrogen bonding network.

From the results obtained using only FTIR spectroscopy, it can be concluded that the electromagnetic field (EMF) influences the dipole moment and hydrogen bonds of the water molecule (Xueyun et al.2016). Therefore, EMF has a direct and strong effect on its chemical structure.

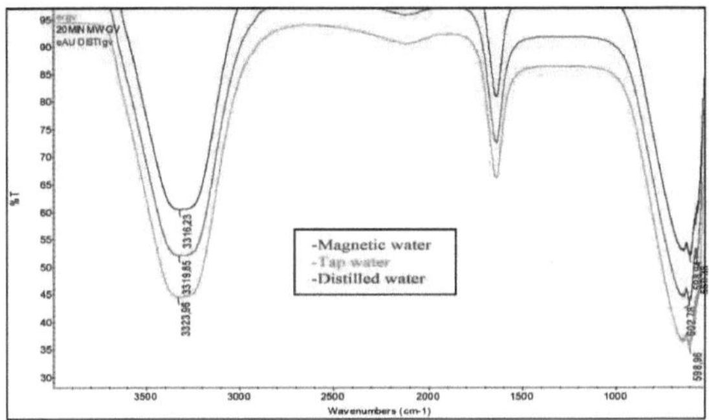

Fig.69. The infrared transmission spectrum of magnetised water affected by the electromagnetic field for 20 min V= 0.6 m s-1, tap water and distilled water.

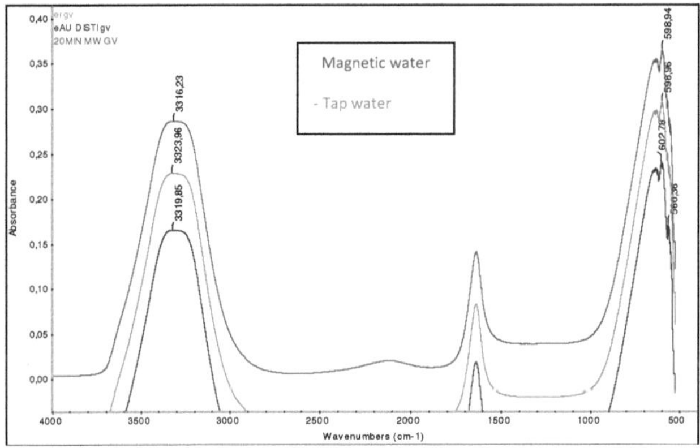

Fig.70. Infrared absorption spectrum of magnetised water affected by the electromagnetic field for 20 min at v= 0.6 m s-1, tap water and distilled water.

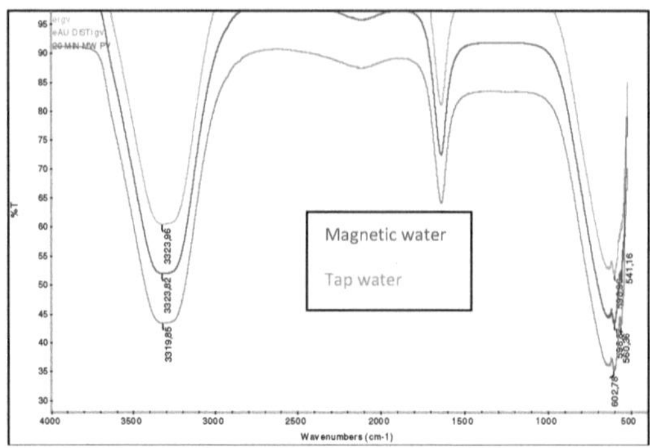

Fig.71. Infrared transmission spectrum of magnetised water affected by the electromagnetic field for 20 min at v= 0.18 m s-1, tap water and distilled water.

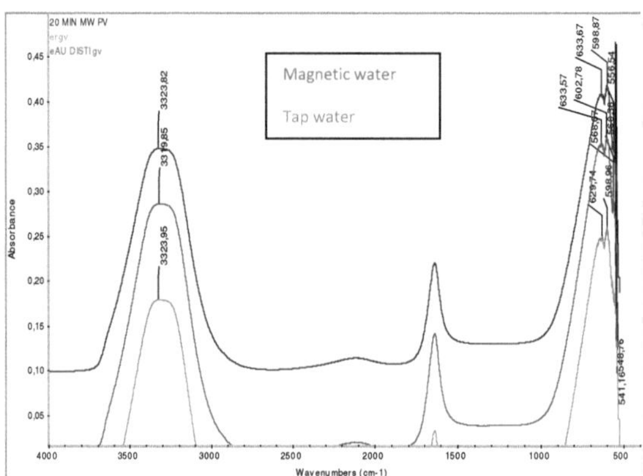

Fig. 72. Infrared absorption spectrum of magnetised water affected by the electromagnetic field for 20 min V= 0.18 m s-1, tap water and distilled water.

ELECTROMAGNETIC MEMORY OF WATER

VII.1 Discussion and Results of the Electromagnetic Memory of Water

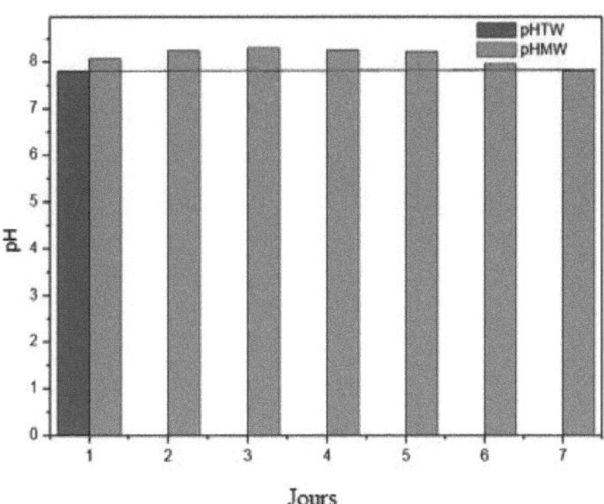

Fig.73. pH of magnetised tap water (water memory) at V = 0.18 m / s and magnetisation time (tm) = 20 min.

Figure 73 shows the pH values of the magnetised tap water as a function of time after the water has been circulated in a closed loop for a period of 20 minutes at a speed of 0.18 m/s through the Aqua 4D device. Before circulation, the pH is 7.8. We deduce that during the 20 minutes of magnetisation, the pH of the water was affected by the recirculation through the Aqua 4D device. The averages of the values measured each day in figure 73 show a decrease of 3.32% of the pH value in 7 days to the initial value of the tap water before magnetisation.

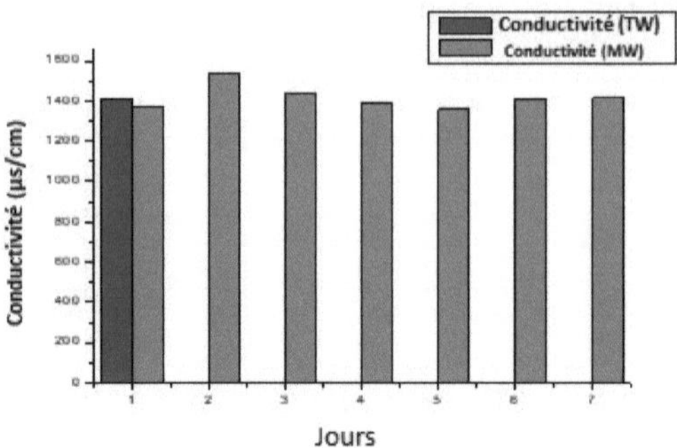

Fig.74. Conductivity of magnetised tap water (water memory) at v = 0.18 m / s and magnetisation time (tm) = 20 min.

Figure 74 shows the conductivity values of magnetised tap water as a function of time after a closed loop circulation of water for a period of 20 minutes at a velocity of 0.18 m / s through the Aqua 4D device. Before circulation, the conductivity value is 1413 µS / cm. We deduce that during the 20 minutes of magnetisation, the conductivity of the water was affected by the recirculation through the Aqua 4D device. The averages of the daily measured values in figure 74 show an increase of 3.66% in the conductivity value of the water in 7 days to the initial value of the tap water before magnetisation.

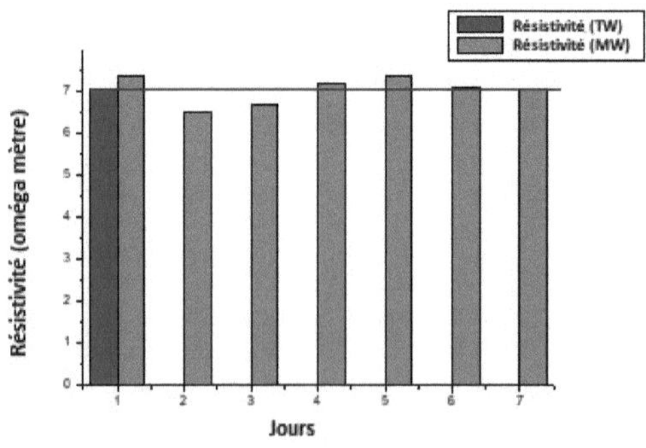

Fig.75. Resistivity of magnetised tap water (water memory) at v = 0.18 m / s and magnetisation time (tm) = 20 min.

Figure 75 shows the resistivity values of magnetised tap water over time after the water has been circulated in a closed loop for a period of 20 min at a rate of 0.18 m / s through the Aqua 4D device. Before circulation, the resistivity value is 7.04 µm. We deduce that during the 20 minute magnetisation, the resistivity of the water was affected by the recirculation through the Aqua 4D instrument. The averages of the daily measured values in Figure 75 show a decrease of 4.1% in the resistivity value in 7 days to the initial value of the tap water before magnetisation.

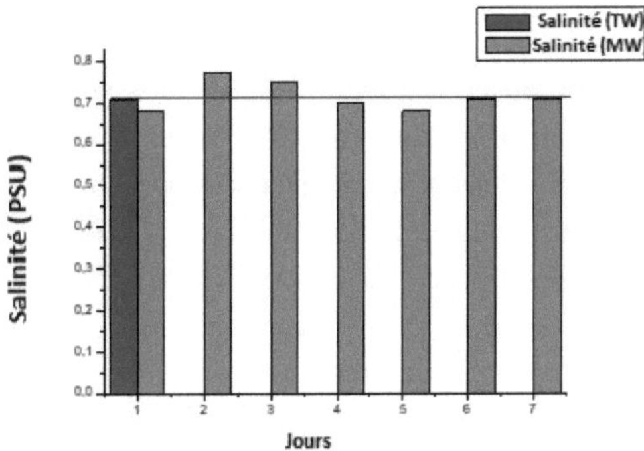

Fig.76.Salinity of magnetised tap water (water memory) at v = 0.18 m / s and magnetisation time (tm) = 20 min.

Figure 66 shows the salinity values of magnetised tap water versus time after the water has been circulated in a closed loop for a period of 20 minutes at a speed of

of 0.18 m / s through the Aqua 4D device. Before circulation, the salinity value is 0.71 PSU. We deduce that during the 20 minutes of magnetisation, the salinity of the water was affected by the recirculation through the Aqua 4D device. The averages of the values measured each day in Figure 76 show an increase of 4.4% in the salinity value in 7 days to the initial value of the tap water before magnetisation.

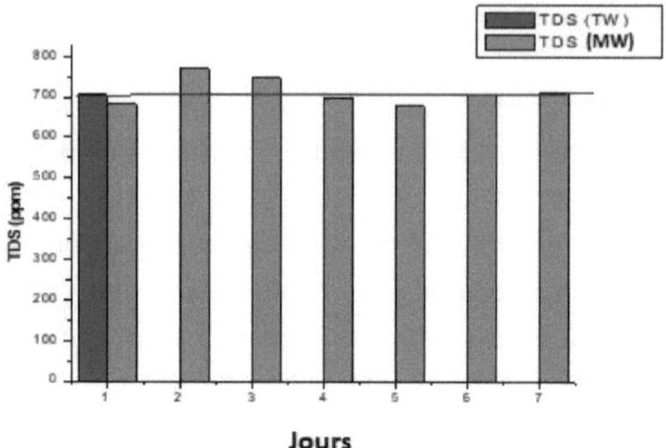

Fig. 77: TDS of magnetised tap water (water memory) at v = 0.18 m / s and magnetisation time (tm) = 20 min.

Figure 77 shows the TDS values of the magnetised tap water as a function of time after the water has been circulated in a closed loop for a period of 20 min at a speed of 0.18 m/s through the Aqua 4D device. Before circulation, the TDS value is 706 ppm. We deduce that during the 20 minutes of magnetisation, the TDS of the water was affected by the recirculation through the Aqua 4D device. The averages of the values measured each day in Figure 77 show an increase of 4% in the TDS value in 7 days to the initial value of the tap water before magnetisation

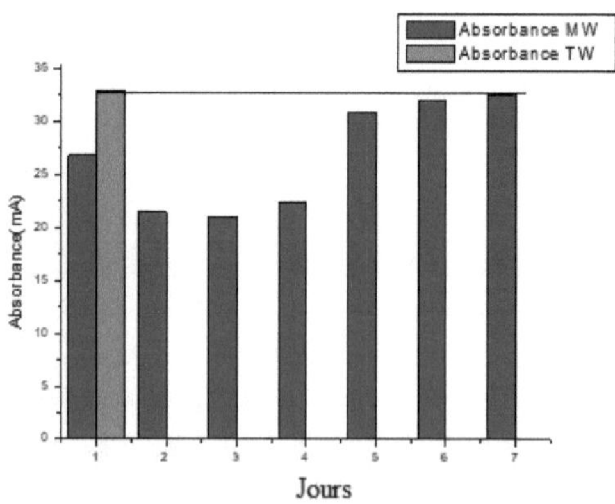

Fig. 78: Absorbance of magnetised tap water (water memory) at v = 0.18 m / s and magnetisation time (tm) = 20 min.

Figure 78 shows the 631 nm absorbance values of the magnetised tap water over time after the water was circulated in a closed loop for a period of 20 minutes at a speed of 0.18 m/s through the device. Aqua 4D. Before circulation, the absorbance value at 631 nm is 32.9 mA. We deduce that during the 20 minutes of magnetisation, the absorbance at 631 nm of the water was affected by the recirculation through the Aqua 4D device. The averages of the daily measured value in figure 78 show an increase of 21.64% of the absorbance value at 631 nm in 7 days to the initial value of the tap water before magnetisation.

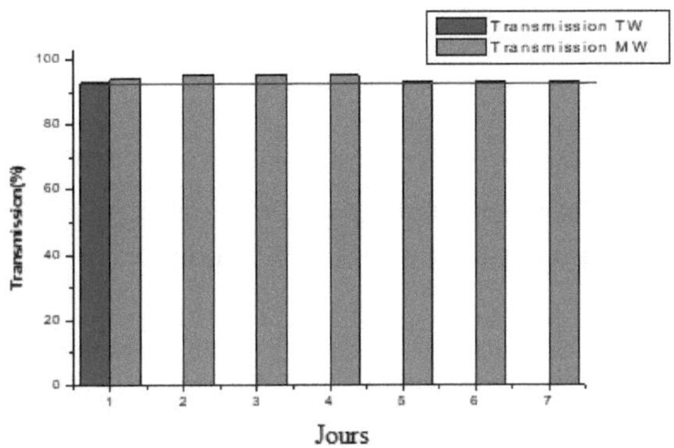

Fig.79 Transmission of magnetised tap water (water memory) at v = 0.18 m / s and magnetisation time (tm) = 20 min.

Figure 79 shows the transmission values of magnetised tap water over time after a closed loop water circulation for a period of 20 minutes at a flow rate of 0.18 m / s through the Aqua 4D device.Before magnetisation the transmission value is 92.72%. We deduce that during the 20 minutes of magnetization, the water transmission was affected by the recirculation through the Aqua 4D device. The average values measured each day in figure 16 show a decrease of 1.32% and the transmission lasts 7 days before returning to the initial value of the tap water before magnetisation.

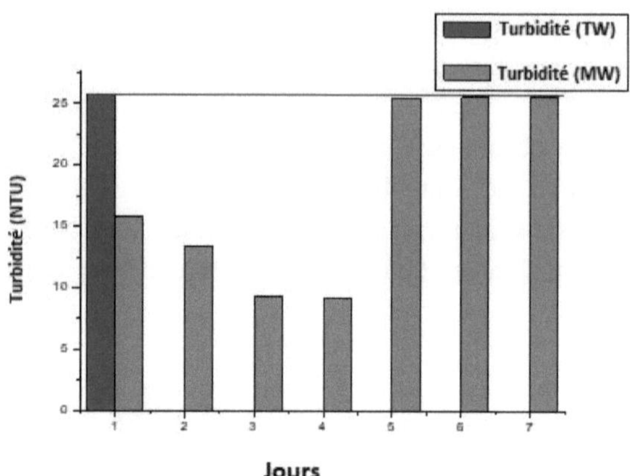

Fig.80.Turbidity of magnetised tap water (water memory) at v = 0.18 m / s and magnetisation time (tm) = 20 min.

Figure 80 shows the turbidity values of the treated water versus time after circulating the water in a closed loop for 20 minutes at a speed of 0.18 m/s through the Aqua 4D device. Before circulation, the turbidity is 25.7 NTU. We deduce that during the 20 minutes of magnetisation, the turbidity of the water was affected by the recirculation through the Aqua 4D device. The averages of the values measured each day in Fig.80 show a 62% increase in turbidity value in 7 days to return to the initial value of the tap water before magnetisation.

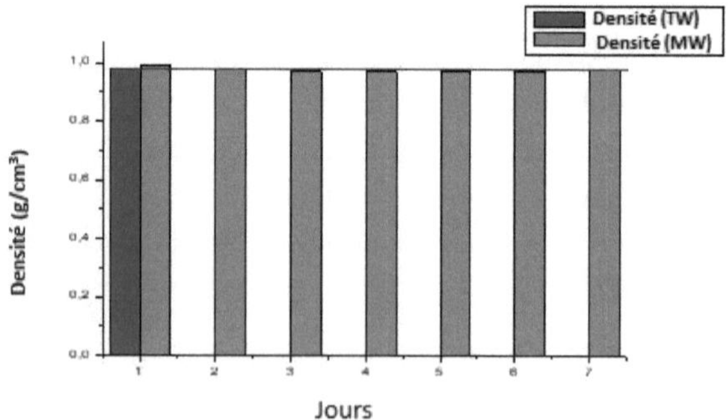

Fig.81.Density of magnetised tap water (water memory) at v = 0.18 m / sand magnetisation time (tm) = 20 min.

Figure 81 shows the density values of the magnetised tap water as a function of time after the water has been circulated in a closed loop for a period of 20 minutes at a speed of 0.18 m/s through the Aqua 4D device. Before circulation, the density value is 25.7 g/cm3. We deduce that during the 20 minutes of magnetisation, the density of the water was affected by the recirculation through the Aqua 4D device. The averages of the values measured each day in figure 81 show a decrease of 2% in the density value of the water in 7 days to the initial value of the tap water before magnetisation.

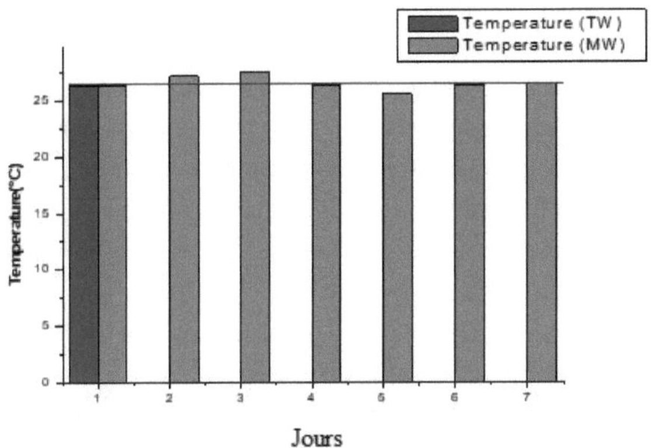

Fig.82: Temperature of magnetised tap water (Memory of Water) at v = 0.18 m / s and magnetisation time (tm) = 20 min.

Figure 82 shows the temperature values of the magnetised tap water as a function of time after a closed loop water circulation for a period of 20 minutes at a flow rate of 0.18 m/s through the Aqua 4D device. Before circulation, the temperature value is 26.4°C. We deduce that during the 20 minutes of magnetisation, the water temperature was affected by the recirculation through the Aqua 4D device. The average values measured daily in figure 82 show an increase of 0.76%.

The temperature to return to the initial value of the tap water before magnetisation within 7 days.

The question that arises here is how much the water retains the values of physico-chemical parameters (pH, salinity, conductivity TDS, absorbance, transmittance, turbidity, temperature, density) obtained by water passing through the electromagnetic field created by Aqua 4D for 20 min after the magnetised water is stored in a static state for seven days. This question illustrates the effect of the electromagnetic field on the memory of the magnetised water. This study can answer the last question; 20 litres of water were magnetised for 20 minutes of recirculation in Aqua 4D, the magnetised water is stored and isolated from atmospheric air to eliminate the effect of absorption of atmospheric gases by the stored water. The timer is restarted and after each day a water sample is taken from the stored water and the values of the physicochemical descriptors are measured each time. The total storage time is 7 days; Figures (73, 74, 75, 76, 77, 78, 79, 80, 81 and 82) show the values of (pH, salinity, conductivity TDS, absorbance, transmittance, turbidity, temperature, density) of the stored water as a function of time. As illustrated in the figures, the water retains and retains the impact of passing through the magnetic field, it is shown to retain the values of the physico-chemical parameters of the magnetised water, and the water memory continues to forget the magnetic impact of these values over a period of 7 days. This phenomenon defines the term (Water Memory) and can be defined as: the moment when the magnetised water remembers the impact of the magnetic field. We can see that magnetised tap water samples slightly increase the rate of magnetism and also the memory effect as the magnetic field is maintained for 7 days indicating the memory effect. It is found in the literature that the memory of water towards a magnetic treatment can last up to 200 hours (Cefalas et al.2010). According to (Cefalas et al.2010), the amplified magnetic mode is not decayed from the forbidden nature of the transition between antisymmetric and symmetric states and "remains trapped in the true volume of all molecular water rotors", which explains the memory effect of water. The magnetic field can be understood from a coherent macroscopic antisymmetric state induced by an external magnetic field on a set of two-level rotors (coherence state of water). The individual molecular rotors are forced to rotate coherently by the external MF. The amplified magnetic mode does not decay into the corresponding coherent symmetric ground state, but remains in the coherent volume of all molecular rotors. The existence of water is antisymmetric. The coherent state explains the memory effects on water (Cefalas et al.2010) by the memory of water.

PHYSICO-CHEMICAL PARAMETERS OF MAGNETISATION

VIII.1 Discussion of Parameters Influencing the Water Magnetisation Process

IX.1.1 Water magnetisation phenomenon

In order to verify the storage capacity of water for electromagnetic energy, we studied the evolution of the magnetic field received during treatment by the electromagnetic field generated by the Aqua-4D device (Figure 83). The linear regression of the fluctuations of the magnetic field received by the tap water shows a decrease of a non-zero value (0.0075 mT) after eight hours (28800 s), which shows the phenomenon of magnetisation of the water. Therefore, water is a substance with the ability to store electromagnetic energy.

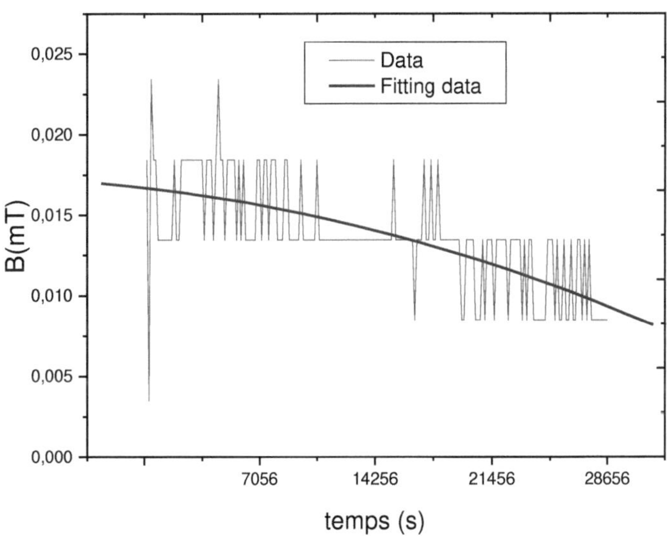

Fig. 83. Monitoring fluctuations in the static magnetic field of magnetised water

During the 15 hours of electromagnetic treatment, the fluctuations of the magnetic field recorded are relatively small (Bmax = 0.0230 mT and Bmin = -0.025 mT) with an average value of 0.0015 mT. This variation is slightly higher than the value measured before the electromagnetic field was applied. These evaluations of the magnetic field flux density allowed us to monitor the possible fluctuations of the magnetic field after exposure of both types of water (magnetised water

(MW) and tap water (TW)) to the electromagnetic field. The experiments indicate that the tap water did indeed become magnetised after treatment with the flux density of the magnetic field generated by the aqua-4D device. The evolution of the static magnetic field after the electromagnetic treatment generated by the aqua-4D device was monitored. The experimental results show that during the 15 hours of electromagnetic treatment, the magnetic field fluctuations recorded are relatively low (Bmax = 0.023 mT and Bmin = 0.008 mT) averaging 0.0155 mT for magnetic water, while for tap water the values obtained are Bmax = 0.0185mT and Bmin = 0.005mT with a Bavg equal to 0.01175mT. The magnetic field measurements allow monitoring of any fluctuations in the field after the samples have been exposed to the electromagnetic field.

IX.1.2 Effect of temperature on electromagnetic field strength and magnetisation time

According to the literature (Amor et al.2018) , the thermal agitation of water causes a rotational motion of the hydrogens (positively charged) around the oxygen nucleus (negatively charged), which forms a short-lived electric current generating a Lorentz force in a uniform static magnetic field that influences the chemical and physical properties of water. In order to better understand the effect of temperature on the magnetisation of water, we varied the temperature and observed the influence of this variation on the magnetisation time of the water. Figures 84 illustrate the influence of temperature on the effect of magnetisation by the electromagnetic field generated by the Aqua 4D device. According to the previous figure, the water passes through an electromagnetic field and becomes magnetised water (MW). The magnetisation times at 26.5°C, 40°C are 15 h, 12h respectively, so there is a 3h difference in Tm between the two temperature conditions. Therefore, the increase in temperature decreases the magnetisation time of water. Concerning the relationship between the magnetic field strength and temperature we observed that the Bavg at 26.5°C is equal to 0.020 mT, but at 40°C the value of Bavg does not exceed a value of 0.002 mT. Thus, the increase in temperature significantly decreases the magnetic field strength. This experimental observation is consistent with the fundamental laws of magnetic moments in electromagnetism. The thermal agitation of water molecules causes disorder by Brownian motion and the disappearance of magnetic order.

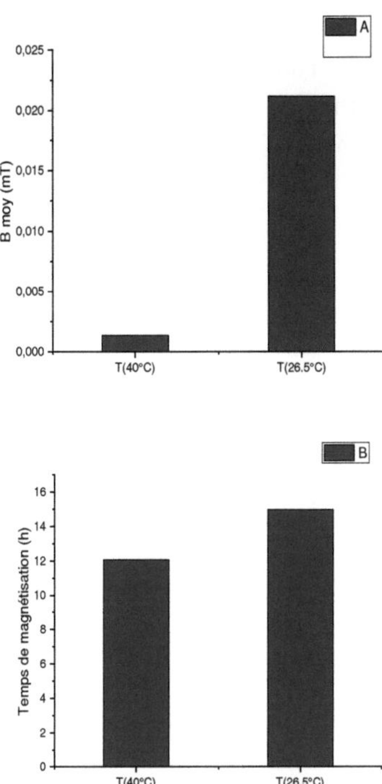

Fig. 84. (A) Average magnetic field strength of water as a function of temperature. (B) Magnetisation time as a function of water temperature.

IX.1.3 Effect of salinity on electromagnetic field strength and magnetisation time

Water is diamagnetic, in the presence of a magnet, the diamagnetism of the water will repel the magnet. When salt is added, it increases the magnetic field strength of the water. So we tried to study this phenomenon. Indeed, a variation of the salinity at a fixed temperature of 26.5°C is carried out, leaving the conductivity of the water to increase under the action of the electromagnetic field. The profile of the electromagnetic field variation with time in the case of salt water compared to tap water indicates that the electromagnetic field is affected by salinity. The impact of salinity on the EMF intensity is very small (Figure 85), indeed the EMF intensity at salinity (0.048 g/l) is Bavg = 0.020 mT, the increase in salinity of 35.952 g/l caused a slight increase in the magnetic field intensity (MFI) of 0.0025 mT therefore, the impact of salinity on the MFI is very small Concerning the magnetization time (Tm) of the water, we observe the decrease of Tm with the salinity, so at a salinity of 0.048 g / l, the magnetization time is equal to 11.5 h, but at a salinity of 36 g / l we obtained 15 h in Tm, so the increase of salinity prolonged the Tm (Figure 85). Thus, the salinity factor favours the magnetisation of the water by lengthening the magnetisation time of the water with little impact on the magnetic field strength.

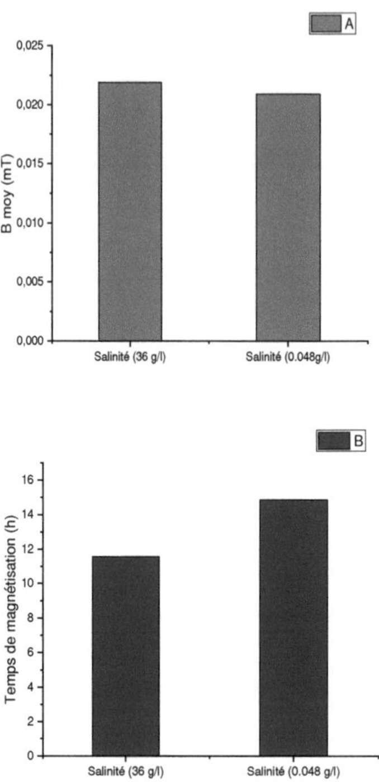

Fig. 85. (A) Average magnetic field strength of water as a function of salinity. (B) Variation of the magnetisation time as a function of water salinity.

IX.1.4 Effect of pH intensity on electromagnetic fields and magnetisation time

In the literature (Alimi et al.2006), some works have studied the effect of the magnetic field on the characteristics of the water, in particular the variation of the pH on treated and untreated water. In our case, we focused on the effect of pH on EMF intensity and magnetisation time. The results are shown in Figure 86.

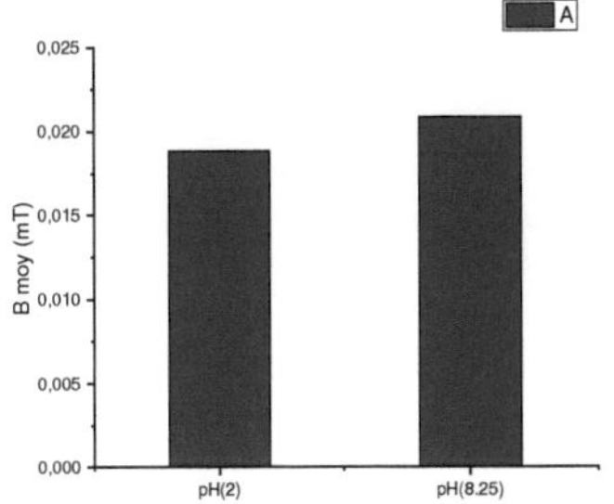

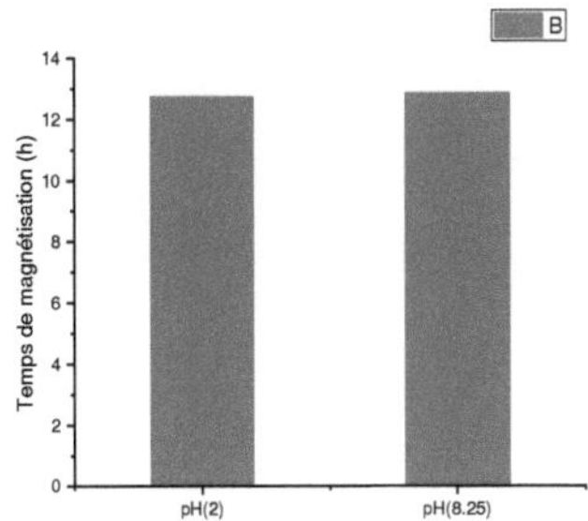

Fig. 86. (A) Average magnetic field strength of water as a function of pH. (B) Magnetisation time and water pH.

According to Figure 86, at pH = 2 (very acidic environment) the magnetic field strength is equal to 0.018 mT. Increasing the acidic pH to basic pH = 8.25, increases the magnetic field strength by 0.002 mT, so the impact of the pH variation on the IMF is small. On the other hand, regarding the impact of pH on the magnetization time of water, the following can be observed. The magnetization time at pH = 2.00 is 15 h, but at pH = 8.25 we obtained a value of 15.30 h as magnetization time, so the influence of pH at the time of magnetization is very small.

IX.1.5 Effect of water velocity on EMF and magnetisation time

Among the most cited parameters in water magnetisation studies, we found the velocity parameter. For example, (Alimi et al.2006) tested the variation of the magnetic treatment efficiency when the flow velocity was changed. In our case, we studied the effect of water velocity on the electromagnetic fields and magnetisation time.

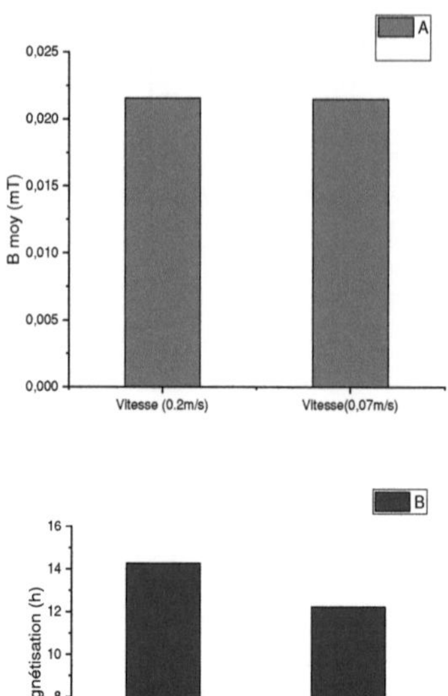

Fig. 87. (A) Average magnetic field strength of water as a function of flow rate. (B) Magnetisation as a function of the two tap water flow rates

From our results, we observed in the figures (Fig. 87) the following. At 0.2 m / s, the EMF intensity is 0.0025 mT but at 0.07 m / s, we obtained 0.0016 mT, therefore, the decrease of water velocity by 0.193 m/s decreases the magnetic field intensity from 0.025 mT at v= 0.2 m /s to 0,0152 mT at 0.07 m / s, therefore, the water velocity influenced the magnetic field intensity of the water. Concerning the impact of the water velocity on the magnetization time we observe that the Tm is equal to 15 h at 0.2 m / s. but at 0.07 m / s we found only 12 h as magnetization time, so the speed of water affects its magnetic field and its magnetization time. Changing the water velocity by 0.13 m/s caused an increase of 0.004 mT in the IMF and extended the Tm by 2 hours.

GERMINATION STUDY OF LETTUCE

IX.1 Effect of the electromagnetic field on the pH of tap water at different speeds

.

The experiment was repeated several times. The results were presented in (figure 88: **(a, b)**).

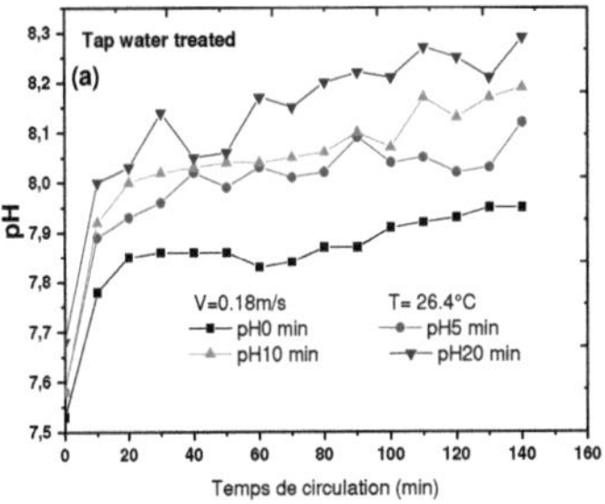

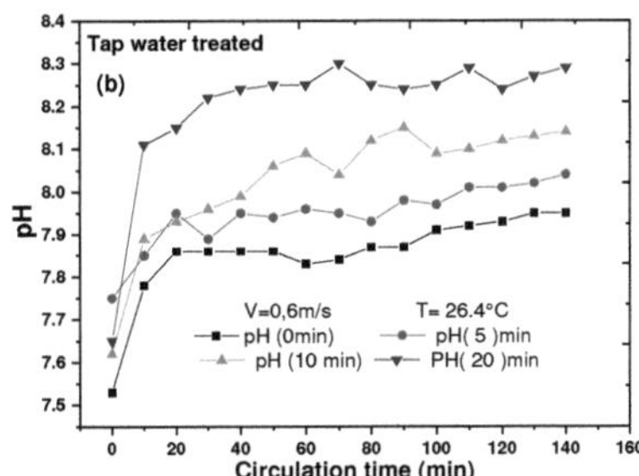

Figure 88: (a, b) pH changes of tap water caused by magnetisation for 5 min, 10 min and 20 min as a function of time, at T = 26.4°C, flow rate v = 0.18 m / s, v = 0.6 m / s respectively (pH at 0 min untreated)

IX.2 Germination study with magnetised water renewed every 24 h at tm = 80 min

The results show that the germination rate was higher for seeds treated with magnetised water renewed every 24 h at tm = 80 min. The rate of germination occurred more rapidly (Figure 89).

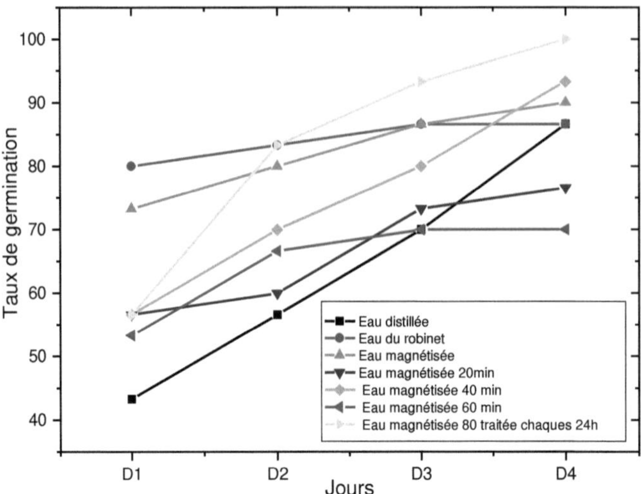

Figure 89: Germination rate of lettuce seeds treated with magnetised water tm = 20 min, tm = 40 min, tm = 60 min, tm = 80 min renewed every 24 h, distilled water and tap water.

The germination percentage of lettuce seeds treated with tap water (TW) and distilled water (DW) is always lower than that corresponding to water treated with the MW electromagnetic field, which modifies the physicochemical properties of the water; the optimal external electromagnetic field can influence the germination rate and percentage (Florez et al. 2007).

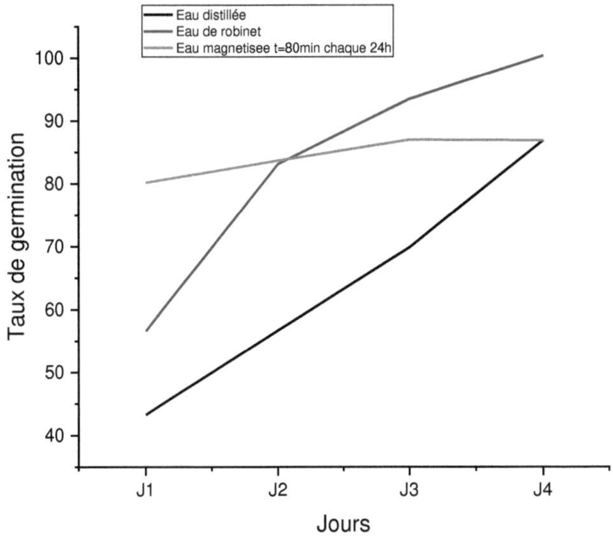

Figure 90: Germination rate of lettuce seeds treated with TW, DW and magnetised water renewed every 24 hours at t = 80min

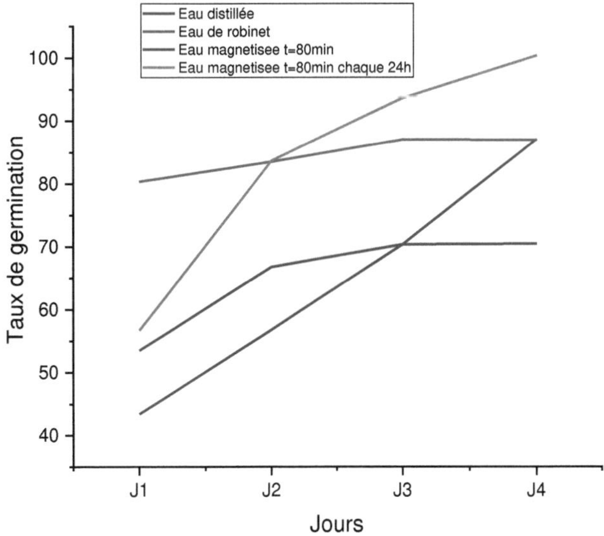

Figure 91: Germination rate of lettuce seeds treated with TW, DW, magnetised water t = 80min and magnetised water renewed every 24 hours at t = 80min

The response of lettuce seeds to magnetised water treated with the electromagnetic field depends on the magnetisation time of the water, the species and the varieties of seeds (Figure 90 and 91).

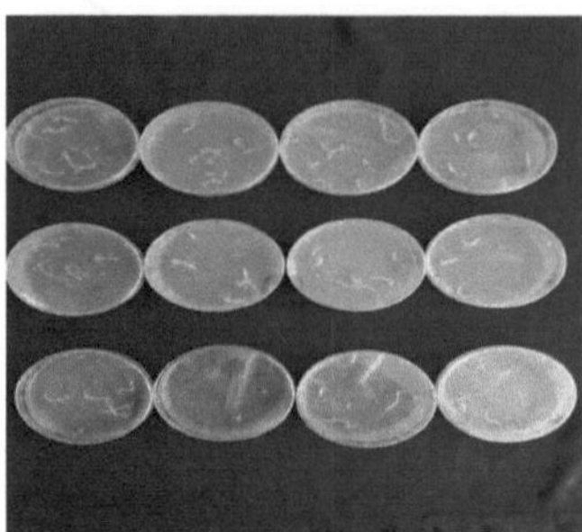

Figure 92: Effect of magnetic treatment on lettuce seed germination tm = 80min with MW renewal every 24h

Seed emergence counts were made with a progression of radicals visible through the seed coat. Data on germination rate and mean germination times were recorded up to 4 days. The germination rate was faster for seeds treated with magnetised water than for untreated seed water. The longest roots and shoots were recorded with the water treated seeds and the shortest stem and roots were observed in the untreated seeds. The main objective of this study was to evaluate the effects of electromagnetic treatment on the germination of milk seeds Fig 92 and 93.

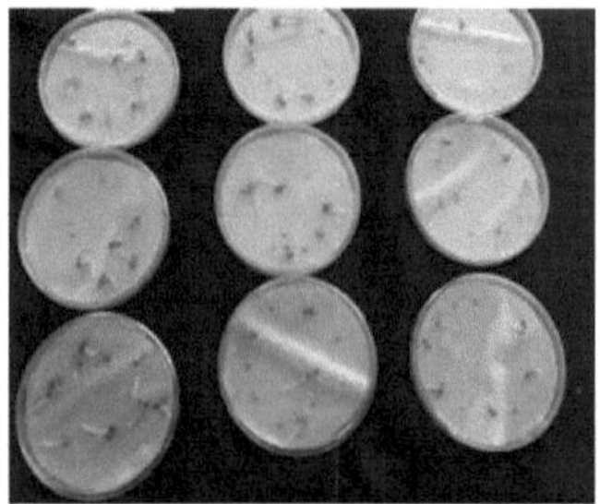

Figure 93: Effect of magnetised water on lettuce sprouts

IX.3 Germination study with magnetised water renewed every 12 h at tm = 80 min

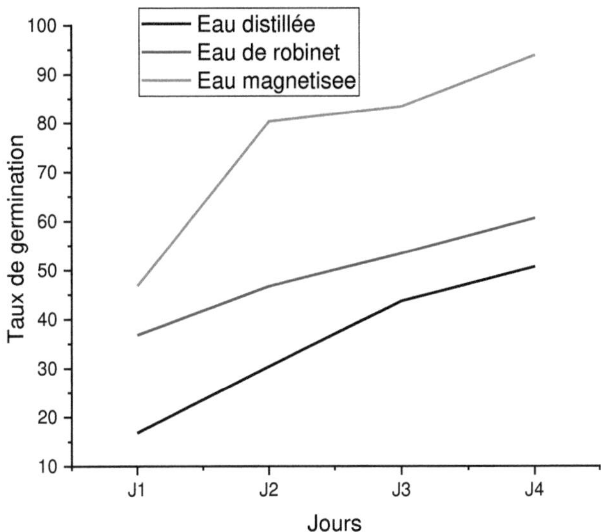

Figure 94: Magnetic treatment on lettuce seed germination tm = 80min with renewal of magnetised water (MW) every 12h

The results show that the germination rate was higher for seeds treated with magnetised water renewed every 12 hours at tm = 80 min. The onset of germination occurred more quickly (Figure 94).

IX.4 Discussion of germination of lettuce (Lactuca sativa L.)

The germination percentage of lettuce impregnated with magnetized water (MW) is higher than that impregnated with tap water (TW) and distilled water (DW), the electromagnetic field has an impact on the properties of water, the rate of change of germination was influenced by the treatment of magnetized water by the electromagnetic field (Cheikh et al., 2018). The germination percentage reached 46.6% after four days of germination of lettuce seeds. It reaches a value of 93.3% with a treatment of treated water (tm = 80 min) (TW) renewed every 12 hours. For tap water, the germination rate increased from 36.6% to 60%. For distilled water, the percentage increases from 16.66% to 50% for lettuce seeds (figure 94). From the results obtained in (Figure 89 and 94), we can conclude that the germination percentage is higher for lettuce seeds treated with magnetized water for 80 min and renewed every 12 h (93.3%) in second position comes the lettuce seeds treated with magnetized water for 80 min and renewed every 24 h (87.5%) so the seed treated with magnetized water for 80 min with four days of renewal is the lowest (70%), which proves that the magnetization of water weakens with time. These results are in agreement with the work of (florez and al.2012) the application of magnetic field increases the germination rate and percentage of sprout treatment compared to untreated plants. According to (Shabrangi et al. 2009). The magnetic field promotes the germination rate of bean and wheat seeds and, in addition, the treated plants grew faster than the control plants (Shabrangi et al. 2009). Furthermore, a positive effect of magnetic treatment on germination and emergence of bean varieties was confirmed; seedling emergence is one of the most important factors in pod production per plant (Podlesny et al., 2004). Many authors have found an increase in growth rate, improved protein synthesis and increased root growth (Carbonell et al., 2000, Florez et al., 2007). Many studies have found a higher percentage of germination and greater plant growth

GROWTH AND ROOT SYSTEM STUDY OF LETTUCE

X.1 Characterisation of tap water and magnetised water

All physical and chemical parameters measured for magnetised water and tap water are collected in Table 5. The results show that the effect of electromagnetic fields on the physico-chemical properties of the water used for lettuce processing is significant.

X1.2 Effect of electromagnetic field on TDS of tap water

Figure 77 shows the TDS values of the treated tap water as a function of time, after it has been circulated in a closed loop for a period of 20 min at a speed of 0.18 m / s through the Aqua 4D device. It can be seen that prior to circulation, the TDS gives a value of 706 ppm. Thus, during the 20 minutes of magnetisation, the TDS of the water was clearly affected by the electromagnetic field generated by the Aqua-4D device. The averages of the values measured each day in figure 77 show an increase in the TDS value of 4% in 7 days, then TDS returned to its initial value of the tap water before magnetisation (memory effect).

X1.3 Calculation of fresh mass of lettuce irrigated with magnetised water every 24 h (= 80 min)

The fresh mass of lettuce irrigated with TW (38 mg) and DW (37, 58 mg) is always lower than that corresponding to the water treated with MW (55, 65 mg). It can be suggested that the electromagnetic field modifies the physico-chemical properties of the water. In fact, some researchers have stated that the optimal external electromagnetic field can influence the fresh mass of lettuce (Moon et al. 2000).

After 24 hours of growth (Fig. 95), the rate of increase in fresh mass of lettuce treated with magnetised water (tm = 80 min) and renewed every 24 hours is 47% higher than with distilled water and 46.4% higher than with tap water.

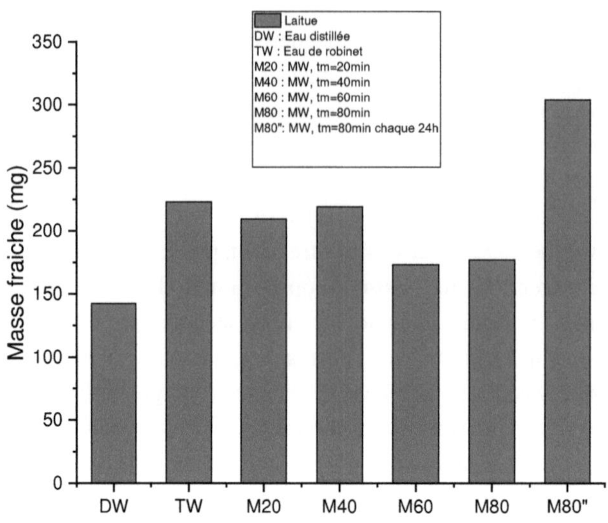

Type of treatment

Fig. 95. Fresh lettuce mass treated with magnetised water (tm = 20 min, tm = 40 min, tm = 60 min, tm = 80 min) renewed every 24 h, and distilled and tap water.

X.I.4 Calculation of the dry mass of lettuce irrigated with magnetised water every 24 h (tm = 80 min)

The dry mass of lettuce treated with TW (2.19 mg) and DW (2.19 mg) is always lower than that corresponding to water treated with MW (2.75 mg) produced by exposure to the electromagnetic field, which changes the physical characteristics of the water. According to the literature, the optimal external electromagnetic field can significantly influence the dry mass of lettuce (Kataria et al. 2017).

After 24 hours of growth (Fig. 96), the rate of increase in the dry mass of lettuce treated with magnetic water exposed for 80 min to the electromagnetic field and renewed every 24 hours was 20.08% higher than with distilled water and 25.57% higher than with tap water.

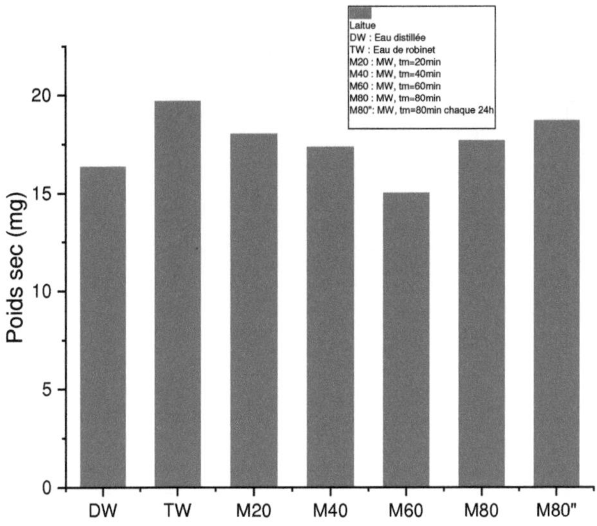

Fig. 96. Dry mass of lettuce treated with magnetised water (tm = 20 min, tm = 40 min, tm = 60 min, tm = 80 min) renewed every 24 h, distilled water and tap water.

X.5 Calculation of root elongation of lettuce irrigated with magnetised water every 24 h (tm = 80 min)

The root elongation of lettuce treated with TW (4.08 cm) and DW (3 cm) is always lower than that observed for water treated with MW (4.27 cm), which modifies certain physical and chemical characteristics of the water.

In previous work, it was reported that the optimal external electromagnetic field could indeed influence the length of lettuce roots (Florez et al. 2007).

After 24 hours of growth, the rate of increase in root elongation treated with magnetic water exposed for 80 min to the electromagnetic field renewed every 24 hours was 47% higher than with distilled water and 46.44% higher than with tap water (Fig.97 and Fig.98).

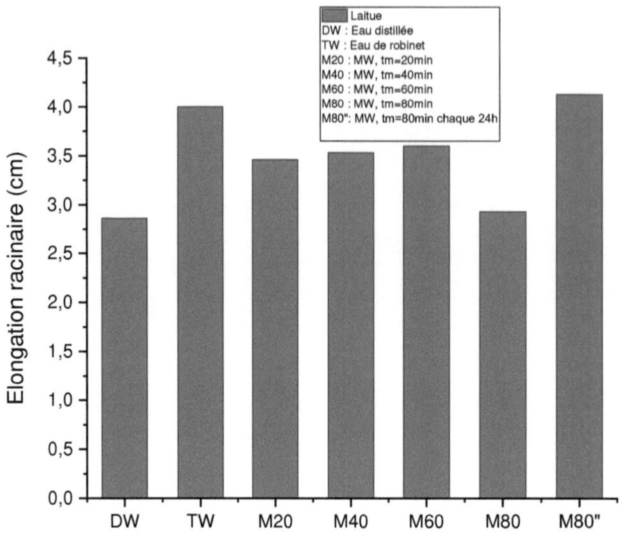

Fig. 97. Root elongation of lettuce irrigated with magnetised water (tm = 20 min, tm = 40 min, tm = 60 min, tm = 80 min) renewed every 24 h, distilled water and tap water.

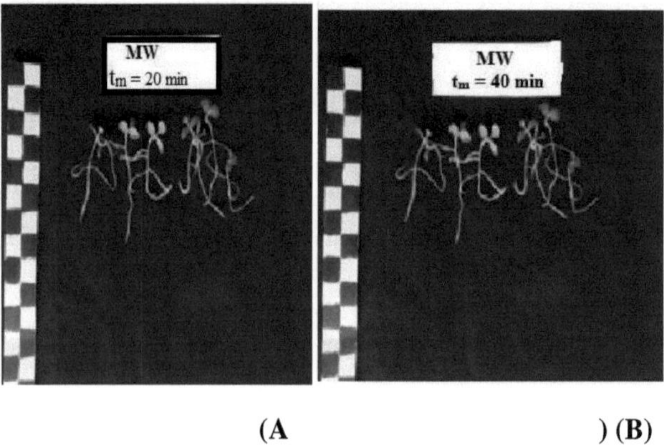

(A) (B)

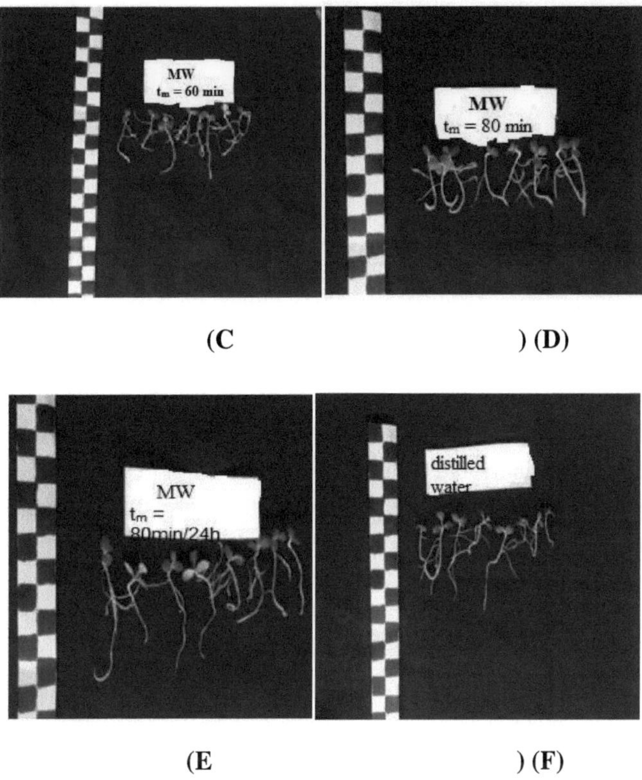

Fig.98. Evolution of the growth of lettuce seeds according to the type of treatment: with magnetised water tm = 20 min (A), tm = 40 min (B), tm = 60 min (C), tm = 80min (D), tm = 80min renewed every 24 h (E), distilled water (F) and tap water (G).

X.5 Discussion

The main reason for the positive effect of using magnetised water is due to the fact that the magnetic field decreases the number of hydrogen bonds and the intensity of Van der Waals forces. As a result, the water will be more cohesive, penetrate more easily and be more easily attached to the roots, which increases the development of the plant (Nakagawa et al. 1999)

. Plants irrigated with magnetised water take up more minerals from the soil and no residues are formed on the soil surface (Abobatta et al .2015, Hachicha et al.2016). The greater penetration of magnetized water than other types through plant cell membranes is explained by the small size of the magnetized water clusters (Silva et al. 2014 , Alimi et al.2009, Raiteri et al.2010 , Colic et al.1998 , Mghaiouini et al.2020, Shiyab et al.2020). Therefore, the penetration of magnetised water through cell membranes, the uptake by the roots will be easier (Barefoot et al. 1992,Kronenberg et al.2005). As a result, root elongation and growth of lettuce were improved (Barefoot et al.1992), which positively influenced the rate of leaf cell division and elongation and the thesis of photosynthesis (Tkachenko et al.1997, Toledo et al.2008, Murad et al.2006, Ben Amor et al.2017, Mghaiouini et al 2020, Shiyab et al.2020). On our side, we have shown that magnetized water has a positive impact on lettuce (LS) growth (plant height, freshness, higher dry mass and root elongation). The stimulating effect of water magnetization on growth criteria may be due to its effect on biochemical changes or altered enzymatic activities, such as ploughing and head stages of some cereal crops, which are affected by the use of magnetized water compared to irrigation with untreated water (Nasher et al.2008, Celik et al. 2008 , Mghaiouini et al .2020 , Shiyab et al.202). Furthermore, the stimulating effect of magnetised water irrigation on growth criteria can probably be due to the induction of mitosis and cell metabolism, since all cat-alytic processes involving oxidation or reduction are accelerated in these plants. This increase and acceleration of growth leads to plant activity and development, related to the increase of gibberellic acid (GA3), ribonucleic acid (RNA), deoxyribonucleic acid (DNA) and enzymatic activities. According to (Qodos et al. 2010), these effects improved the growth criteria of the plant. The improvement in root elongation suggests that electromagnetic treatment appears to be an effective method that can be used in the field or in the greenhouse where improved root growth will allow better moisture uptake. Improved root elongation can lead to earlier growth of lettuce (LS) and therefore faster uptake of mineral nutrients and water, allowing higher yields.

XI.1 Mechanical study of mortars

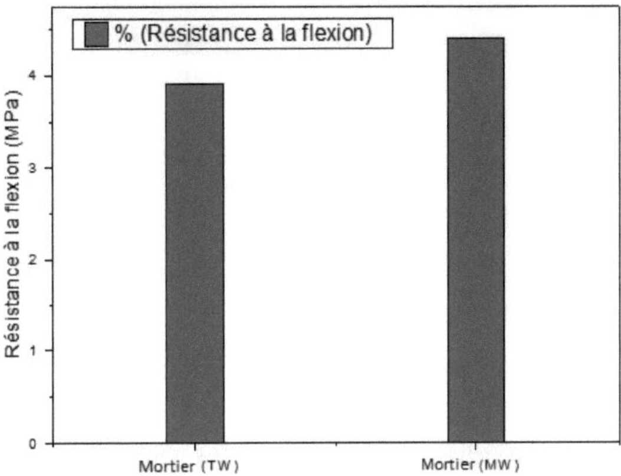

Figure 99. Flexural strength of normal mortar prepared with tap water (TW) and magnetised water (MW)

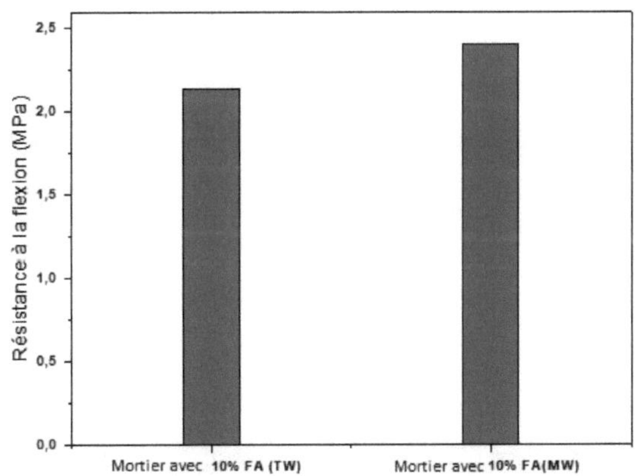

Figure 100: Flexural strength of mortar prepared with 10% fly ash (FA) with tap water (TW) and mortar with 10% fly ash (FA) and magnetised water (MW)

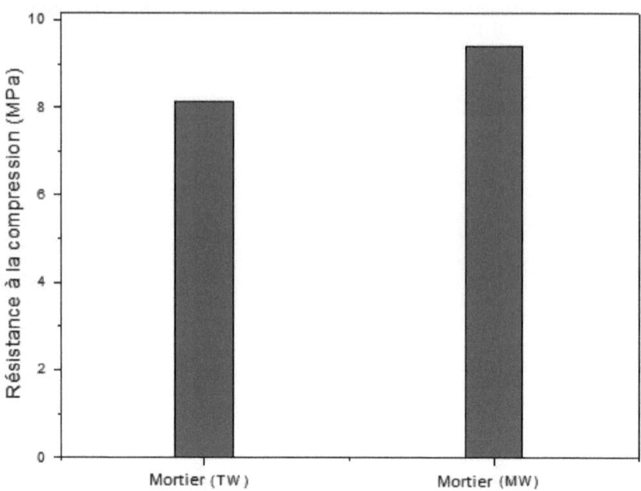

Figure 101. Compressive strength of mortar prepared with 10% fly ash (FA) with tap water (TW) and mortar with 10% fly ash (FA) and magnetised water (MW)

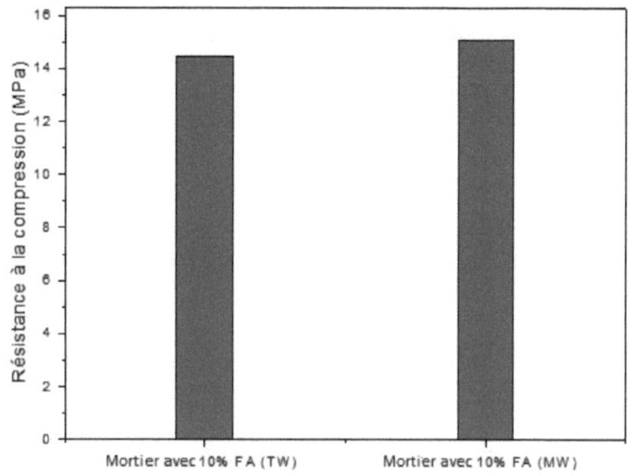

Figure 102. Compressive strength of mortar prepared with 10% fly ash (FA) with tap water (TW) and mortar with 10% fly ash (FA) and magnetised water (MW)

Figures 99 and 100 show the flexural strength of the 28 day cured mortar samples. Figures 101 and 102 show the relative compressive strength of the 28-day mortar samples. Although prepared with varying percentages of cement substitution by fly ash, all mortar samples show a similar trend, indicating that the effect of the EMFTW magnetic field strength on the compressive strength of different mortar samples is almost identical. As can be seen, a percentage of fly ash is used instead of cement, the compressive strength and flexural strength of the samples mixed with EMFTW are higher than those of the control (tap water being represented by 0 T). In other words, EMFTW is more effective than tap water during the hydration process. Within the electromagnetic field generated by AQUA 4 D, the magnetic force can break up small clusters of water into clusters, which increases the efficiency of the water (Aarfane et al.2014) . As the hydration of cement grains does, the distribution and penetration rate of WFETs through the almost impermeable layer of cement paste is higher than that of tap water. As a result, hydration is improved, which increases the strength of the mortar. Furthermore, during hydration, the magnetised water clusters become smaller and disperse and penetrate easily through the cement particles, which improves the reaction of water with cement and the interface of magnetised water with sand and fly ash particles will increase (Yu et al. 1998). In addition, mortar made by WPCM contains fewer micropores because it is denser and therefore less permeable to water through the mortar (Su et al. 2000).

XI.2 Characterisation of mortars by X-ray spectroscopy

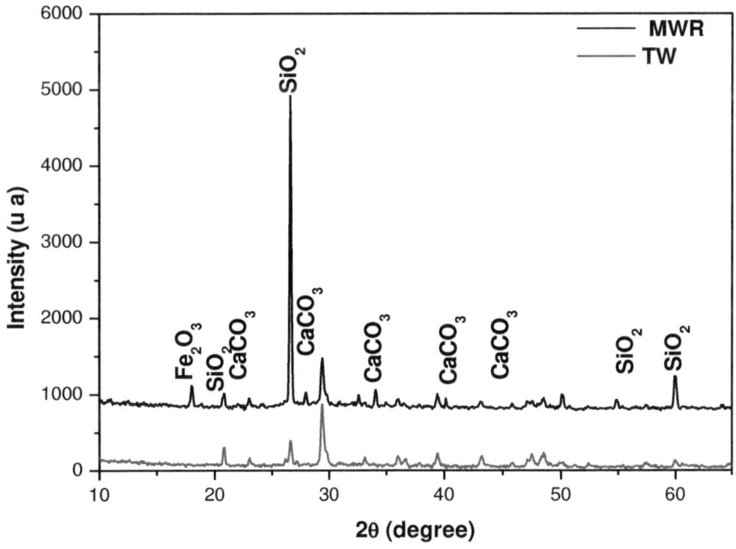

Figure 103. X-ray diffraction pattern of mortars with TW and MW

From the X-ray diffraction pattern in Figure 103, three main phases were identified, namely the calcium carbonate (CaCO3) phase which was observed at the (2 theta) positions (23.01°; 29.37°; 35.98°; 39.41°; 43.14°). Another phase identifies that of silicon oxide (SiO2) which was revealed at the (2 theta) positions (20.79°; 26.60°; 39.41°; 40.17°; 45.83°; 50.04°; 59.93°; 68.07°) and finally the phase of iron oxide (Fe2O3) at the (2 theta) position (18.03°)

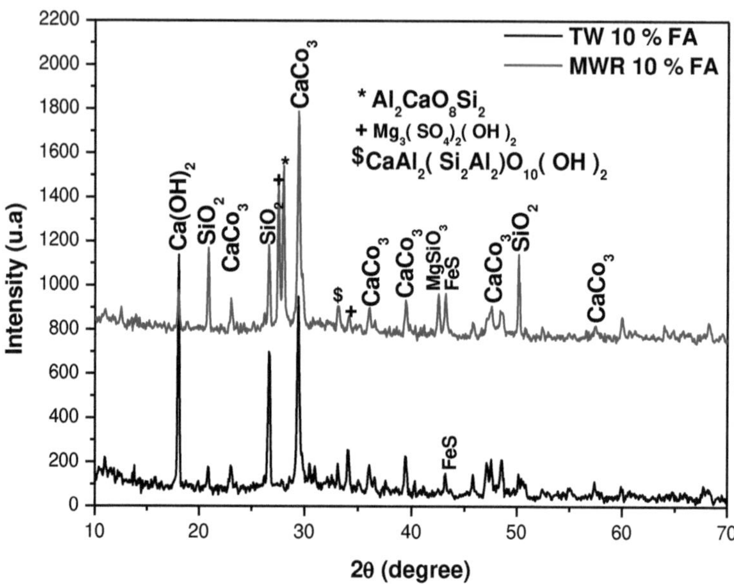

Figure 104. X-ray diffraction pattern of mortar with TW with FA and MW with FA

From the X-ray diffraction pattern in Figure 104, three main phases were identified, namely the calcium carbonate (CaCO3) phase which was observed at the (2 theta) positions (23.04°; 29.41°; 36.02°; 39.44°; 47.45° ; 57.42°), another phase identifies that of silicon oxide (SiO2) which was revealed at the (2 theta) positions (20.85°; 26.63°; 50.15°) two new phases are also observed in the mortars prepared with (MW) which are the Mg3 $(SO4)_2(OH)_2$ phases at the (2 theta) positions (27.5°, 34.1°) and the Al2CaO8SiO4 phase at the (2 theta) positions (28.04°).

3.3.1 Interpretation of X-ray and crystallite size results in mortars

Figures 103 and 104 show that the most marked differences between the four types of mortars consist in the creation of new crystalline phases in the mortars prepared with (MW) which are the $Mg3(SO4)_2 (OH)_2$ phase at the (2 theta) positions (27,5°,34,1°) and the Al2CaO8SiO4 phase at the (2 theta) positions (280,04°), which proves the existence of a crystallisation process on the one hand and the influence of the electromagnetic field strength on the other hand on the chemical hydration process of the cement. It can also be seen in figure 103 that the most intense XRD peak is that of silicon oxide (SiO2), whose content increases 13 times compared to the peak (SiO2) of the mortar prepared by tap water due to the influence of the electromagnetic field on the mortar preparation water, which is in agreement with (Madsen et al. 2004) . Figure 103 also shows that the peak height of calcium carbonate (CaCO3) (5.1 cps) in the mortar prepared with (MW) is large compared to that of the mortar prepared with (TW) (4.3 cps) something which shows that magnetised (MW) water promotes the crystallisation of calcium carbonate. This is consistent with the other characterisation results presented above regarding physical, mechanical and microstructural properties. In a previous paper, it was concluded that a magnetic field accelerates the crystallisation of poorly soluble diamagnetic salts of weak acids and calcium carbonate (Saddam et al. 2014). It is also observed in Figure 104 that the peak height of the CSH phase of calcium silicate hydrate in (MW) based mortar is 17.5% higher than that of (TW) based mortar, the CSH phase is the important phase i on which the physical characteristics and more particularly the mechanical characteristics of the mortar evolve (Joshi et al. 1966). In another figure 104, the peak of portlandite $Ca (OH)_2$ in fly ash mortar and (TW) is three times higher than the peak of fly ash mortars and (MW) because the water magnet favours the reaction of portlandite with the mortar components which contributed to the reduction of the peak intensity in the mortar based on the water magnet (MW) but this phase is the low contribution on the mechanical strength on the other hand, portlandite is important in the durability mechanism (Joshi et al. 1966). The smaller water molecules of MWFET reacted with the mortars. This difference explains why the compressive and flexural strength of cementitious materials with EMFTW is higher than that with tap water.

XI.3 Effects of MFTW on the microstructure of mortars

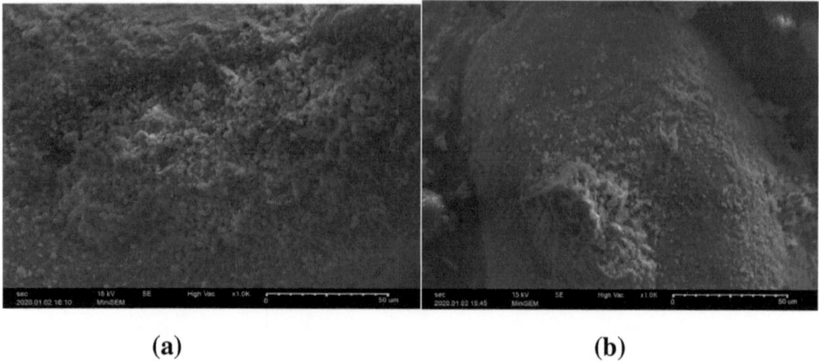

(a) (b)

Figure 105. SEM micrograph of a mortar prepared with MW and TW (1000 ×).

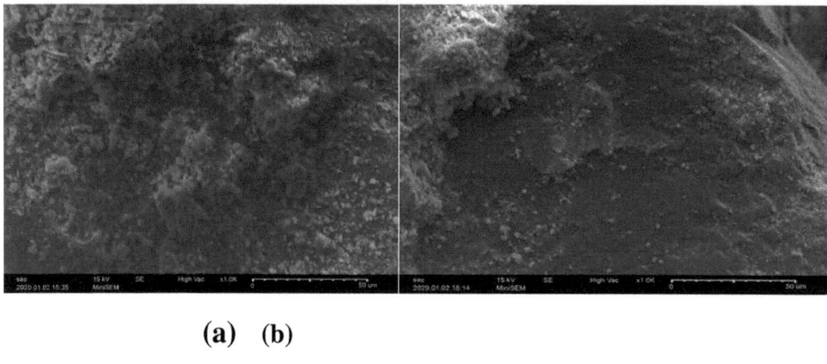

(a) (b)

Figure 106. SEM micrograph of a mortar prepared with 10% fly ash using MW magnetic water and TW tap water (1000 ×).

XII.3 The effects of MFTW on the microstructure of mortars

In order to visualise the morphologies of the cement matrix and fly ash in mortars and composites, SEM photographs (Figures 105 and 106) were taken of mortar samples with and without fly ash. Figures 105 and 106 show cement crystals prepared with tap water and MFTW, respectively. The morphology of the cement in the mortars prepared with magnetic water is close to that prepared with tap water. The tap water molecules tend to clump together and form clusters, which could be produced after the cement reacts with these clumped water molecules (Figure 105a and Figure 106a). Figures 105b and 106b show that the cement crystals in the MW mortar tend to be more dispersed than the mortars prepared with TW tap water and form separately, the compressive and flexural strength of the MW mortars is higher than that with tap water because the cluster size of MFTW is smaller reacted with the cement.

FORMULATION AND CHARACTERISATION OF FIRE-ASH BASED MORTAR

XII.1 Mechanical characterisation of mortars

Figures 107 and 108 illustrate respectively the mechanical behaviour of flexural and compressive strengths of mortars prepared with tap water (TW) and magnetised water (MW) after 28 days of curing. Figures 109 and 110 explain experimentally other mechanical parameters of the mortar which are the flexural and compressive strengths after 28 days of curing of the samples containing 10% BA mixed with tap water (TW) and magnetised water (MW). Although prepared with varying percentages of cement bottom ash substitution, all mortar specimens show a similar evolution, indicating that the effect of the magnetic field strength of the TWMF on the compressive strength of different mortar specimens is almost identical. As can be seen, if bottom ash is used instead of cement and regardless of the bottom ash content, the flexural and compressive strength of the mortar mixed with EMFTW is higher than those of the control (tap water being represented by 0 T). In other words, EMFTW is more effective than tap water during the hydration process. In addition, the effect of EMFTW on compressive strength varies with the percentage of ash added. The greatest growth is observed when working with 10% bottom ash

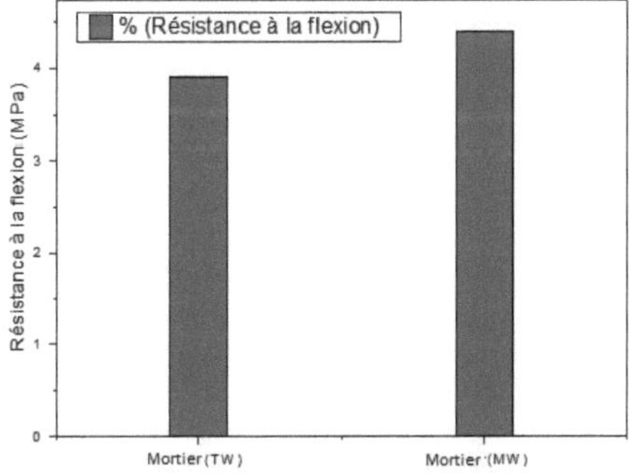

Fig.107. Flexural strength of normal mortar prepared with tap water (TW) and magnetised water (MW)

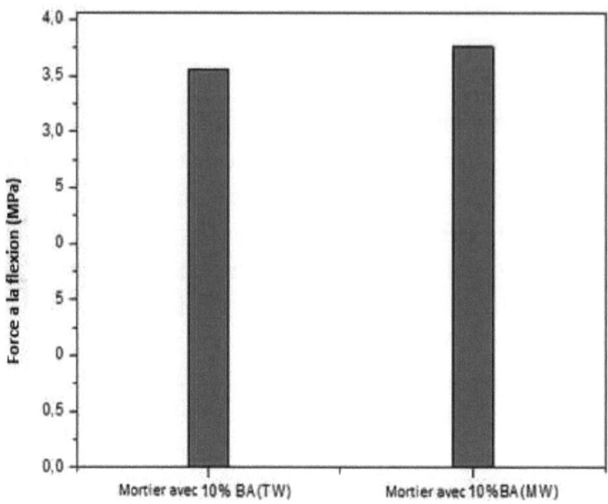

Fig.108. Compressive strength of normal mortar prepared with tap water (TW) and magnetised water (MW)

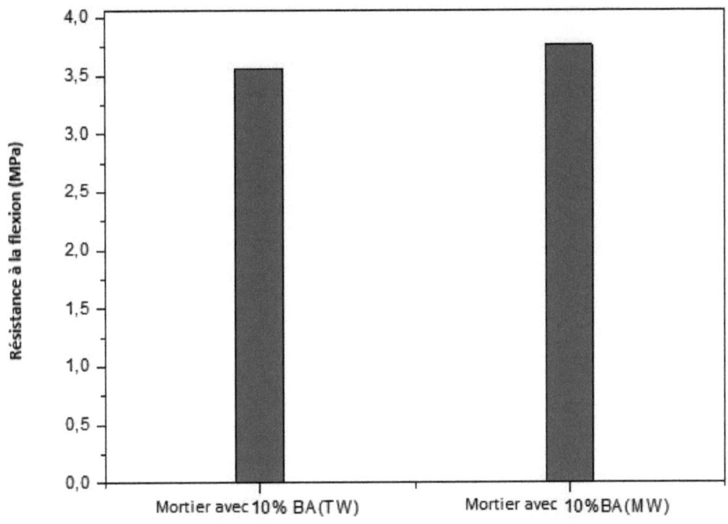

Fig.109. Flexural strength of mortar prepared with BA 10%(TW) and BA 10%(MW)

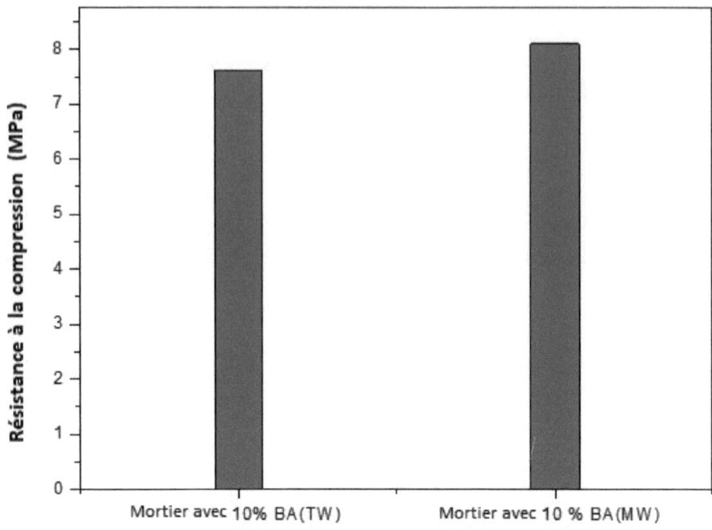

Fig.110. Compressive strength of mortar prepared with BA 10%(TW) and BA 10%(MW)

Figures 108 and 110 show the compressive strength of mortar samples cured for 28 days. Figures 107 and 109 show the relative flexural strength of mortar samples at 7 days. Despite being prepared with different percentages of cement substitution by fly ash, all mortar samples show a similar trend, indicating that the effect of the magnetic field strength of the WFCM on the compressive strength of different mortar samples is almost identical. As can be seen, if fly ash is used instead of cement and regardless of the fly ash content, the compressive and flexural strength of the samples mixed with EMFTW is higher than that of the control (tap water being represented by 0 T). In other words, EMFTW is more effective than tap water during the hydration process. In addition, the effect of WLE on compressive strength varies with the percentage of fly ash. The most significant increase is observed when the fly ash percentage is 10%.

Within the electromagnetic field generated by the AQUA 4 D, the magnetic force can break up water clusters into smaller ones. As a result, the water efficiency is increased (Su et al. 1999). As the hydration of cement grains is taking place, the distribution and penetration speed of EMFTW through the almost impermeable layer of the cement paste is greater than tap water. As a result, hydration is improved, which increases the strength of the mortar. In addition, during hydration the magnetized water clusters become smaller and disperse and penetrate easily through the cement particles which improves the reaction of water with cement and also the magnetized water interface with the sand and fly ash particles will increase (Juan et al.2015, Abdel-Magid et al.2017). In addition, the mortar made by EMFTW contains fewer micro pores as it is denser and therefore less permeability of water through the mortar (An-Tai et al.2007).

XII.2 Characterisation of mortars by X-ray spectroscopy

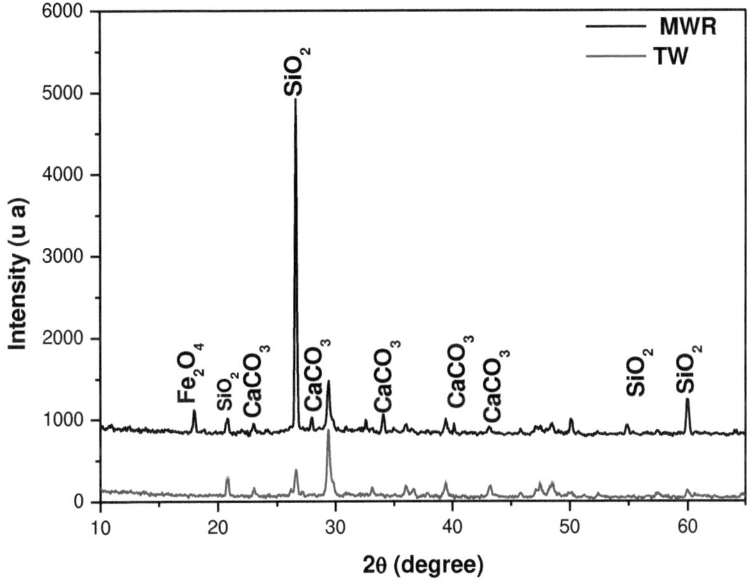

Fig.111. X-ray diffraction pattern of mortars with tap water and magnetised water

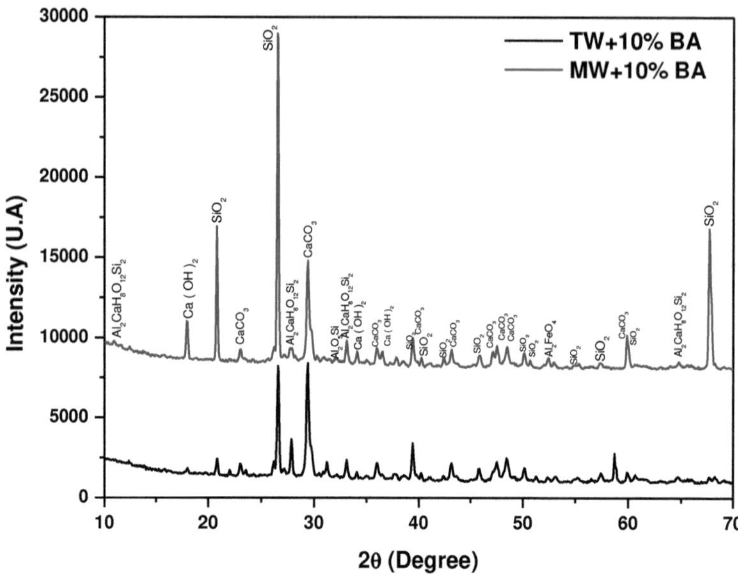

Fig.112. X-ray diffraction pattern of mortars with tap water with 10% machver and magnetised water with 10% machver

From the X-ray diffraction pattern in Figure 111 three main phases were identified which are the calcium carbonate (CaCO3) phase which was observed at the (2 theta) positions (23.01°; 29.37°; 35.98°; 39.41°;43.14°;47.28°; 48.49°), another identified phase which is that of silicon oxide (SiO2) which was revealed at the (2 theta) positions (20.79°; 26.60°; 39.41°; 40.17°; 45.83°; 50.04°; 59.93°; 68.07°) and finally the iron oxide (Fe2O3) phase at the (2 theta) position (18.03°).

From the X-ray diffraction pattern in Figure 112 three main phases were identified which are the calcium carbonate (CaCO3) phase which was observed at the (2 theta) positions (23.04°; 29.41°; 36.02°; 39.44°; 47.45°; 57.42°), another identified phase which is that of silicon oxide (SiO2) which was revealed at the (2 theta) positions (20.85° ; 26.63° ; 50.15°) we also observe two new phases in the mortar prepared with (MW) which are the Mg3(SO4)$_2$ (OH)$_2$ phase at the (2 theta) positions (27.5°, 34.1°) and the Al2CaO8SiO4 phase at the (2 theta) positions (28 .04°) .

XII.2.1 Interpretation of X-ray results

Figures 111 and 112 show that the most marked differences between the four types of mortars consist in the creation of new crystalline phases in the mortars prepared with (MW) which are the $Mg3(SO4)_2$ $(OH)_2$ phase at the (2 theta) positions (27.5°, 34.1°) and the Al2CaO8SiO4 phase at the (2 theta) positions (28 .04°) which proves the existence of crystallization processes on the one hand and the influence of the electromagnetic field strength on the other hand on the chemical reactions of the cement hydration. We can also observe in figure 111 that the most intense peak in figure 111 of XRD is that of silicon oxide (SiO2) whose content increased by thirteen times compared to the peak (SiO2) of the mortar elaborated by tap water due to the influence of the electromagnetic field on the mortar preparation water which is in agreement with (Fargas et al.2001, Asokan et al.2005, Su et al.2000) . Figure 112 also shows that the peak height of calcium carbonate (CaCO3) (5.1 cps) in the mortar prepared with (MW) is large compared to that of mortar prepared with (TW) (4.3 cps) which proves that the magnetised water (MW) promotes the crystallisation of calcium carbonate. This is consistent with other characterisation results presented previously regarding physical, mechanical and microstructural properties. In a previous paper, it was concluded that a magnetic field accelerates the crystallisation of poorly soluble diamagnetic salts of weak acids and calcium carbonate (Fargas et al.2001). We also observe in figure 112 that the peak height of the calcium silicate hydrate C-S-H phase in the (MW) mortar is 17.5% higher than in the (TW) mortar. The C-S-H phase is the important phase on which the evolution of the physical characteristics and more particularly the mechanical properties of the material depend (Haldun Kurama et al.2007).In addition, Figure 112 shows that the peak of portlandite $Ca(OH)_2$ in fly ash and (TW) mortar is three times larger than the peak of fly ash and (TW) mortar but this phase has a small contribution to the mechanical strength but portlandite is important in the durability mechanism (Haldun Kurama et al. 2007).

XII. 3 Interpretation of IR results

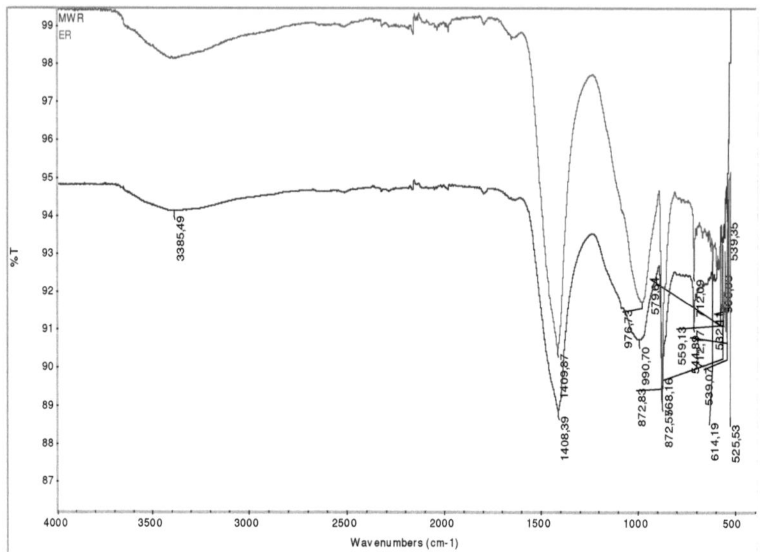

Fig.113. Infrared spectroscopy spectrum of mortars with TW and MW

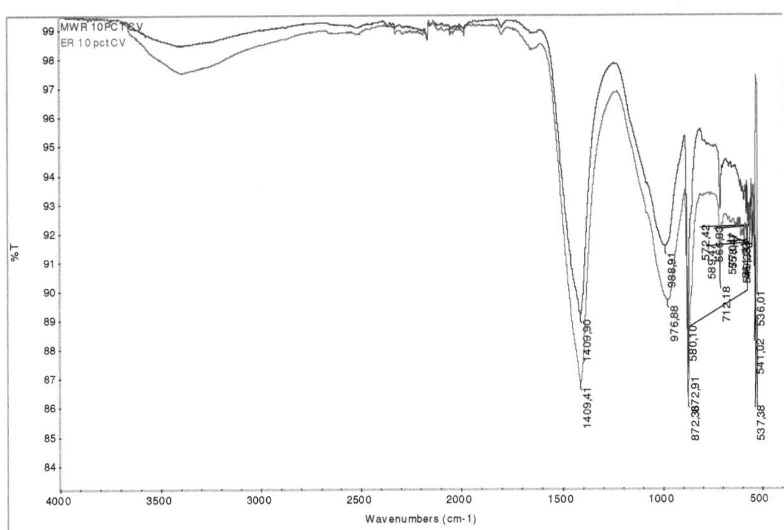

Fig.114. Infrared spectroscopy spectrum of 10% slag mortars with TW and MW

From figures 113 and 114 we observe a band around 525.53 cm-1 (Si-O-Si) which correspond respectively to the valence and deformation vibration modes of the Si-O (silica phase) C_3S and C_2S bond, moreover an intense peak around 976 .73 cm-1 shows a very precise cliniker percentage.We reveal a bonde at 875 cm-1, followed by a changeable peak at 872 .83 cm-1 followed by a peak that changes at 1409 .87 cm-1 characteristic of the elongation vibration modes of the C-O bond. These absorption peaks are characteristic of the existence of calcium carbonate (Ca CO_3) possibly formed by a carbonation reaction of clinker, on the contrary the characteristic bands of the aluminate phase are invisible in the infrared spectrum.

An analysis of the infrared spectra in figures 113 and 114 shows the elongation of the S-O bond of sulphates (SO_4^{2-}) at around 614 cm-1 revealing the existence of gypsum ($CaSO_4$), on the other hand the peaks revealing water (H_2O) appear above 3000 cm-1 (at around 3386cm-1) and a peak relating to a mode of vibration of elongations of the O-H bond

In figure 114 the revelation of the hearth ash (pozzolan) present in small quantities in the composite analysed and which is determined by the vibration of the Si-O bond of the silica at 1030 cm-1 (intense and wide) this component is completely camouflaged by the clinker which makes the detection of this hearth ash phase by the IR technique very deficient (Fargas et al.2001).

XV. Laboratory-scale extraction results

Table 10 and Figures 115 and 116 show that when extracted by a series of KOH solutions (0.1N, 0.5N and 1N) at room temperature for varying times.

The same behaviour is observed when extracting total humic compounds, humic acids and fulvic acids using a solution of KOH (0.5N) for two hours of extractions with magnetised water.

Table 10

Extraction time	Type of water	%OM (g/100g compost) dry weight of compost = 0 .25g			% C in CHT (g/100g compost) dry weight of compost =3.27g		% C in AH (g/100g compost) dry weight of compost =3.27g		% C in AF (g/100g compost) dry weight of compost= 3.27g	
		1N	0, 5N	0,1N	1N	0, 5N	1N	0, 5N	1N	0, 5N
1H	Tap water	17. 5	19.2	10.15	3.41	12.92	0.26	7.72	3.1 5	5.2
	Magnetised water	18.24	20. 67	12.1 6	12.82	13. 66	6.83	8.45	5.99	5.75
2H	Tap water	20.1	23.05	13.1	13	15.0 7	0.43	7.12	12. 57	7.95
	Magnetised water	23.2	24.32	14.2	13.8	15.28	7.19	9.3	6.61	5.98
3H	Tap water	21.7	22. 5	11.2	12. 5	14. 55	0.23	7.9	12.27	6. 6 5
	Magnetised water	22.8	23.1	12.1	12.9	15.0 6	6.99	8.81	5.91	6.25
4H	Tap water	21.14	22.91	11.12	12.31	14.11	0.44	6.71	11.87	7.4
	Magnetised water	22 .3	23.92	12.1 6	11.82	14.8	7.46	8. 63	4. 36	6.17
6H	Tap water	20.2	21.3	10.2	11.1	13.32	0.31	6.1	10.79	7.22
	Magnetised water	21.1	22.92	11.45	10.81	13.81	6.05	8.1	4.7 6	5.71
12H	Tap water	9.12	19.33	7.21	6.11	8.2	0.21	5 .4	5.9	2.8
	Magnetised water	10.13	20.27	8.1	5.82	9.1	5.22	7. 5	0. 6	1. 6
24H	Tap water	8.31	10. 52	7.25	5. 6 2	6.2	0.15	4.9	5.47	1.3

	Magnetised water	10.4	18.3	9.11	4.9 5	7.9	4.49	6.99	0.4 6	0.91

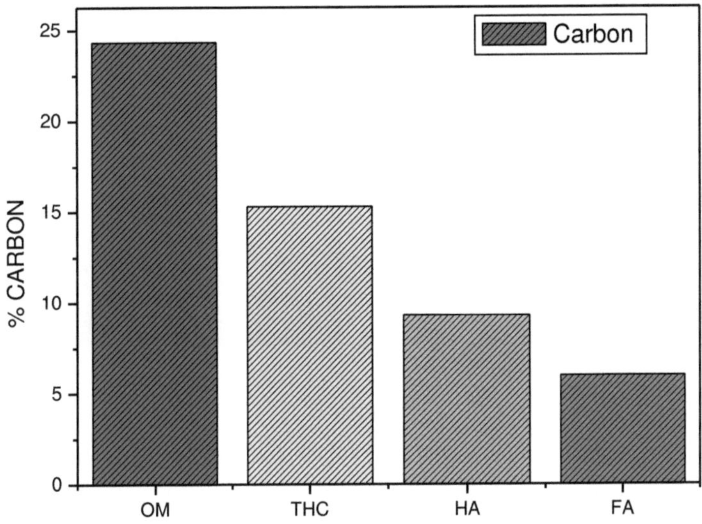

Figure 115: percentages of organic matter (OM), total humic compounds (THC), humic acids (HA), fulvic acids (FA) after 2h of extraction with water magnetized with 0.5N KOH solution

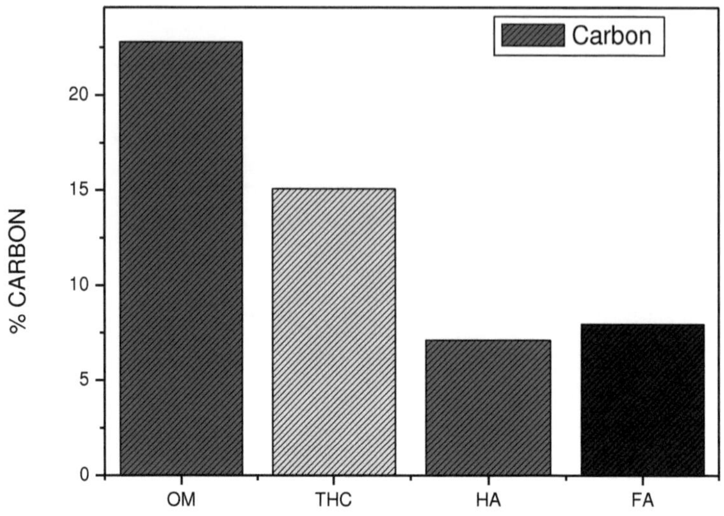

Figure 116 Percentages of organic matter (OM), total humic compounds (THC), humic acids (HA), fulvic acids (FA) after 2 hours of extraction with tap water and 0.5N KOH solution

XVI. Pilot scale extraction results

Table N°11: Extraction of OM from HA and AF by magnetized water from 5kg of compost

Extraction time	Type of water	%MO	%CHT	%AH	%AF
1h	Magnetised water	26	14. 6%	11.08	3.52
2h	Magnetised water	34.8	19.14%	14.88	4.26

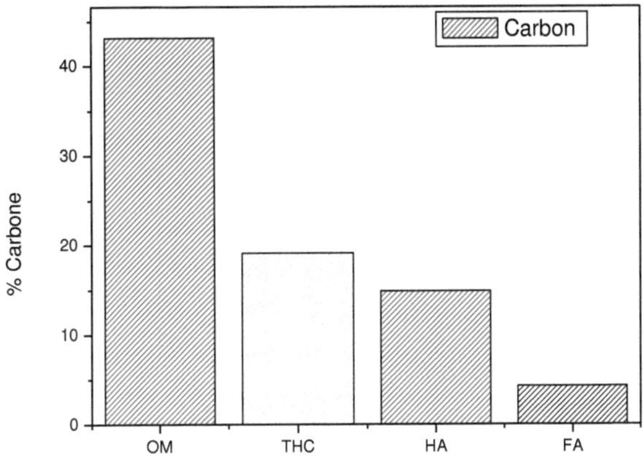

Figure 117: percentages of organic matter (OM), total humic compounds (THC), humic acids (HA), fulvic acids (FA) after 2h of extraction by magnetized water with 0. 5N KOH solution at pilot scale

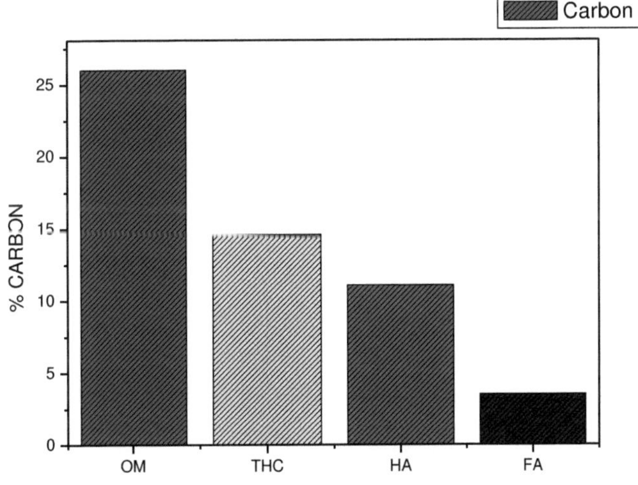

Figure 118: percentages of organic matter (OM), total humic compounds (THC), humic acids (HA), fulvic acids (FA) after 2h extraction with tap water and 0.5N KOH solution at pilot scale

Table 11 and figures 117 and 118 show that when working with a larger quantity of compost, the yields of extractions with magnetised water of OM from HA and AF are significantly improved compared to tap water.

Conclusion

The work presented in this thesis has sought to make as rigorous a contribution as possible to demonstrating the impact of physical treatment of water by a static electromagnetic field on the physicochemical properties of water. It is the potential physicochemical effects of electromagnetic fields, reported in the literature, that motivated us to start this research work.

In general, the effect of electromagnetic field treatment on the physico-chemical properties of water has always generated a lot of controversy due to two facts: - The first is that most of the processes have been patented and are based on commercial interests, which mixes information with promotional claims of beneficial properties that go beyond scientific objectivity.

- The second is the difficulty of understanding the physico-chemical phenomenon that takes place and of understanding the mechanisms that are hypothetical.

The electromagnetic field affects the pH of tap water. This effect can be interpreted by the energy absorbed during the applied electromagnetic field through the dipole moment. The latter causes the water molecule to dissociate under the influence of an electromagnetic field, releasing more OH- ions and absorbing H^+ ions, which causes the pH to rise. The pH increases from 7.6 to 8.25 and by 8.5%. This result is more significant than that reported in other works.The effects of magnetisation speed and time are also demonstrated experimentally. The application of the electromagnetic field during the circulation of water in industrial devices seems to be a promising method in the industrial sector by an economically attractive method. The production of alkaline water becomes easy in practice by applying the electromagnetic field to tap water of constant volume and constant temperature.

In this study, the effects of the electromagnetic field (EMF) on the partial physical properties of water are reported. The characteristics of tap water (TW) and magnetised water (MW) were measured under the same conditions. It was found that the properties of TW were changed as a result of the treatment (CEM), involving an increase in the amount of evaporation and a decrease in the specific heat and boiling point after magnetisation. These changes are highly dependent on the magnetisation effect. In addition, the electromagnetic field strength (EMF) has a marked influence on the magnetisation effect. The results provided an easy approach to improve the cooling and power generation efficiency in industrial devices, which seems to be a promising method to save the energy needed to evaporate water. Furthermore, the present study highlights the validity of magnetisation monitoring using electrochemical impedance spectroscopy (EIS) with distinct dielectric properties, and high and low dielectric constant, respectively. EIS is capable of detecting small traces of water magnetisation that cannot be detected by other techniques.

The electromagnetic field has an effect on the physical-chemical and electrical properties of water. This effect can be interpreted by the energy absorbed during the applied electromagnetic field through the dipole moment. The infrared spectrum gives us a lot of information about the changes in the optical properties of magnetised water compared to other types of water, namely distilled water and tap water, which are taken as controls in this experiment. We noted after an examination of the infrared spectrum that the electromagnetic field influences the intensity and width of the alcohol band and its peak shifts slightly towards lower wave numbers. Both observations can be explained by the influence of the electromagnetic field which directs the electric dipoles of the water molecules in one direction. This rearrangement modifies the angle of the water molecules and consequently a rupture of the hydrogen bonds. These results are very interesting to explain a little bit the phenomenon of magnetisation of water

Water has the ability to store electromagnetic energy. Increasing salinity causes a very slight change in magnetic field strength and extends the magnetisation time by three hours. Thermal agitation of the water decreases the magnetic field strength and extends the magnetisation time by three hours. Varying the pH has a slight effect on the magnetic field strength but not on the magnetisation time. Varying the water velocity affects the magnetic field of the water and the magnetisation time. Changing the water velocity by 0.13 m/s caused an increase of 0.004 mT and lengthened the magnetisation time by 2 hours.

In this thesis we have shown that water retains and retains the impact of the magnetic field crossing for seven days because water has a memory with physico-chemical descriptors measured for seven days as an indicator of water memory.

In addition, the germination capacity and germination rate of lettuce can be improved by the magnetised water produced by the Aqua 4D device.Furthermore, the germination percentage and germination rate of the control lettuce seeds are also lower than those of the applied electromagnetic treatments. In conclusion, the use of electromagnetic field could be an effective solution to improve the germination rate of lettuce.

Our experimental results proved that the application of the electromagnetic field generated by the electromagnetic slide improves the growth parameters of lettuce namely dry mass, fresh mass and root elongation. The treatment with electromagnetic water led to an increase in the percentage of fresh mass, dry mass and root elongation compared to the control for the different types of treatment, respectively. In conclusion, the use of electromagnetic field could be an effective solution to improve the growth and production of lettuce.

Another application in the field of civil engineering is the study of the possible uses of electromagnetic fields for the improvement of the physical-chemical and mechanical characteristics of certain construction materials such as mortar, and to find out the possible processes of the influence of electromagnetic fields on mortars. The tests carried out under the effect of the electromagnetic field showed its influence on the physical characteristics related to the strength of cement mortar based on fly ash waste.
Regarding the flexural strength of the normal mortar without the addition of fly ash prepared with magnetised water, it increased by 12.56% compared to the mortar prepared with tap water and for the mortar with prepared with 10% fly ash and magnetised water the flexural strength increased by 12.67% compared to the mortar prepared with 10% fly ash and tap water. The compressive strength of the normal mortar without fly ash prepared with magnetised water increased by 15.8% compared to the mortar prepared with tap water, and for the mortar prepared with 10% fly ash and magnetised water the compressive strength increased by 4.35% compared to the mortar prepared with 10% fly ash and tap water. The recycling of waste from the thermal power station avoids the depletion of raw materials in nature and takes into account the protection of the environment. It must, of course, be combined with an increased global demand.
We have also tried to valorise the fire-ash which is a waste product of thermal power plants in the field of construction materials in order to replace part of the cement with this environmentally harmful waste. We have experimentally proven that the introduction of magnetised water improves the mechanical properties of normal and fire-ash based mortars.

The flexural strength of the normal mortar formulated with 0% magnetised ash and water was improved by 12.56% compared to the mortar mixed with tap water and also for the mortar prepared with 10% fireplace ash the magnetised flexural strength of the water is improved by 12.67% compared to the mortar mixed with 10% fireplace ash and tap water.

The compressive strength of the normal mortar formulated with 0% hearth ash and magnetised water was improved by 5.91% compared to the mortar mixed with tap water. Similarly, the compressive strength of the mortar prepared with 10% hearth ash and magnetised water was improved by 6.42% compared to the mortar mixed with 10% hearth ash and tap water.The addition of fireplace ash in the formulation of an eco-mortar seems to be a solution that takes into account both the growing environmental, economic and global demand for building materials.

Finally we tried to identify the optimal conditions that can be used to extract organic matter, humic acids and fulvic acids from compost made from bovine manure and coffee grounds. The optimal conditions determined for extracting humic and fulvic acids from compost are: extraction by KOH(0.5 N) for two hours with magnetised water at room temperature. These conditions were extrapolated to the pilot scale with very good reproducibility.

In conclusion, from the above, we can conclude that the electromagnetic field improves the physicochemical properties of water. Magnetised water has many applications in various fields such as industry, agriculture, civil engineering and chemistry.

XVI **Perspectives**

Future experiments are to be carried out with the main objective of solving some of the problems faced by the industry, namely the problem of corrosion of the installations in the plants and also the problem of lime precipitation in the pipes and boilers. Another important aspect is to increase the yield of animal production via magnetised water. In the field of civil engineering we intend to carry out experiments on concrete and bricks with the aim of improving the mechanical properties of these two key materials in the field of construction. After the in vitro test we intend to carry out plantations in the field to improve the plant production by irrigation with magnetised water.

REFERENCES

Alimi, F, Tlili, M, Amor, M, Gabrielli, C,Maurin,G, 2006, Influence of magnetic field
on calcium carbonate precipitation, *Desalination 206 (2007) 163-168*

Lin, i. Yotvat, j, 1990, exposure of irrigation and drinking
Water to a magnetic field with controlled power and
Direction, *journal of magnetism and magnetic materials 83 (1990) 525-526 525 North-holland*

Yue, Y, Huaxiang, W, Wenhui, X, Xiehe, Z, Dingxiang, M, Tingjie, M, SU, L, 1983,
Studies on the effect of magnetized water in the treatment
Of urinary stone and salivary calculus. *Proceedings of the Seventh International Workshop on Rare Earth-Cobalt Permanent Magnets and Their Applications.*

Busch, K.W., Busch, M.A., 1997. Laborator studies on magnetic water treatment and
their relationship to a possible mechanism for scale reduction. *Desalination 109 (2),
131–148.*

Cho, Y, Lee, S, 2005, Reduction in the surface tension of water due to physical water
treatment for fouling control in heat exchangers, *International Communications in Heat
and Mass Transfer 32 (2005) 1*

Nan, S , Yeong-Hwa, W, Chung- Yo, M, 2000, Effect of magnetic water on the engineering properties of concrete containing granulated blast-furnace slag, *Cement
and Concrete Research, v 30, n 4, p 599-605, April*

Krauter, P, Harrar, J , Orloff, K, Bahowick, S, 1996, *Test of a Magnetic Device for the
Amelioration of Scale Formation at Treatment Facility D*, Livermore, California

Lower S, 2009, *Water Cluster Quackery* , CHEM1, viewed 20 April 2009,
http://www.chem1.com/CQ/clusqk.html

Magnetic Water Explained, 2007 , SCALE FIGHTER , viewed 25 Aug 2009, http://www.scalefighter.com/index.php?pr=MWT_Explained

Brower, D, 2005, Magnetic Water Treatment, *Pollution Engineering*

Gholizadeh M., Arabshahi H., Saeidi M. & Mahdavi B., 2008. The effect of magnetic water on growth and quality improvement of poultry. *Middle-East J. Sci. Res.* **3**(3), 140-144.

McMahon C.A., 2009. Investigation of the quality of water treated by magnetic fields.

Bourget S., 2011. Evaluation of the effect of static magnetic fields on post-harvest ripening and senescence of tomatoes.

Cullity B.D. & Graham C.D., 2011. *Introduction to magnetic materials*, John Wiley & Sons.

Vallée P., 2004. Study of the effect of low frequency electromagnetic fields on the physico-chemical properties of water.
General introduction
31

Marshutz S., 1996. Art of scale reduction. *Reeves J* **76**(2), 3.

Ambashta R.D. & Sillanpää M., 2010. Water purification using magnetic assistance: a review. *J. Hazard. Mater.* **180**(1), 38–49.

Marshutz S., 1996. Art of scale reduction. *Reeves J* **76**(2), 3.

Maffei M.E., 2014. Magnetic field effects on plant growth, development, and evolution. *Front. Plant Sci.* **5**, 445.

Otsuka I. & Ozeki S., 2006. Does magnetic treatment of water change its properties? *J. Phys. Chem. B* **110**(4), 1509–1512.

Ali Y., Samaneh R. & Kavakebian F., 2014. Applications of Magnetic Water Technology in Farming and Agriculture Development: A Review of Recent Advances. *Curr. World Environ.* **9**(3), 695–703.

Teixeira J., 1999. Water, liquid or deliquescent crystal? *Research* **324**, 36-39.

Su N. & Wu C.-F., 2003. Effect of magnetic field treated water on mortar and concrete containing fly ash. *Cem. Concr. Compos.* **25**(7), 681–688.

Ibrahim I., 2006. Biophysical properties of magnetized distilled water. *Egypt J Sol* **29**, 363-369.

Smirnov J., 2003. BioMagnetic hydrology. The Effect of a Specially Modified Electromagnetic Field on the Molecular Structure of Liquid Water. Global Quantec. *Inc USA* 122-125.

Pang X.-F. & Deng B., 2008b. The changes of macroscopic features and microscopic structures of water under influence of magnetic field. *Phys. B Condens. Matter* **403**(19), 3571-3577.

Hao X., Liu H., Zhang G., Zou H., Zhang Y., Zhou M. & Gu Y., 2012. Magnetic field assisted adsorption of methyl blue onto organo-bentonite. *Appl. Clay Sci.* **55**, 177-180.

Yavuz H. & Celebi S.S., 2000. Effects of magnetic field on activity of activated sludge in wastewater treatment. *Enzyme Microb. Technol.* **26**(1), 22–27.

Tai C.Y., Wu C.-K. & Chang M.-C., 2008. Effects of magnetic field on the crystallization of CaCO 3 using permanent magnets. *Chem. Eng. Sci.* **63**(23), 5606-5612.

Tai C.Y., Chang M.-C., Liu C.-C. & Wang S.S.-S., 2014. Growth of calcite seeds in a magnetized environment. *J. Cryst. Growth* **389**, 5-11.

Deng B. & Pang X., 2007. Variations of optic properties of water under action of static magnetic field. *Chin. Sci. Bull.* **52**(23), 3179–3182.

Pang X. & Deng B., 2008a. Investigation of changes in properties of water under the action of a magnetic field. *Sci. China Ser. G Phys. Mech. Astron.* **51**(11), 1621–1632.

Lebkowska M., 1991. Effect of constant magnetic field on the biodegradability of organic compounds. *Wars. Univ. Technol. Publ. House Wars. Eff. Constant Magn. Field* 53

Krzemieniewski M., Dobrzynska A., Janczukowicz W., Pesta J. & Zielinski M., 2002. Effect of constant magnetic field on the process of generating hydroxyl radicals wreakcji Fenton. *Chemist* **1**, 12.

Lee S.H., Jeon S.I., Kim Y.S. & Lee S.K., 2013. Changes in the electrical conductivity, infrared absorption, and surface tension of partially-degassed and magnetically-treated water. *J. Mol. Liq.* **187**, 230–237.

Madsen H.L., 1995. Influence of magnetic field on the precipitation of some inorganic salts. *J. Cryst. Growth* **152**(1), 94-100.

Hosoda H., Mori H., Sogoshi N., Nagasawa A. & Nakabayashi S., 2004. Refractive indices of water and aqueous electrolyte solutions under high magnetic fields. *J. Phys. Chem. A* **108**(9), 1461–1464.

Yu Q., Sugita S., Sawayama K. & Isojima Y., 1998. Effect of electron water curing and electron charging curing on concrete strength. *Cem. Concr. Res.* **28**(9), 1201-1208.

Xantheas S.S., 2000. Cooperativity and hydrogen bonding network in water clusters. *Chem. Phys.* **258**(2), 225-231

Moosavi F. & Gholizadeh M., 2014. Magnetic effects on the solvent properties investigated by molecular dynamics simulation. *J. Magn. Magn. Mater.* **354**, 239–247.

Moosavi F. & Gholizadeh M., 2014. Magnetic effects on the solvent properties investigated by molecular dynamics simulation. *J. Magn. Magn. Mater.* **354**, 239–247.

Chang K.-T. & Weng C.-I., 2006. The effect of an external magnetic field on the structure of liquid water using molecular dynamics simulation. *J. Appl. Phys.* **100**(4), 43917.

Toledo E.J., Ramalho T.C. & Magriotis Z.M., 2008. Influence of magnetic field on physical-chemical properties of the liquid water: insights from experimental and theoretical models. *J. Mol. Struct.* **888**(1), 409–415.

Zhou K., Lu G., Zhou Q., Song J., Jiang S. & Xia H., 2000. Monte Carlo simulation of liquid water in a magnetic field. *J. Appl. Phys.* **88**(4), 1802-1805.

Jeon S., Kim D.-R. & Lee S., 2001. Changes of solubility speed of salts in magnetized water and crystal patterns of NaCl, KCl and gypsom intermediated by magnetized water. *J.-KOREAN Chem. Soc.* **45**(2), 116-130.

Lee S.-H., Lee S.-K. & Jeon S.-I., 2009. Study on the Critical Micelle Concentration Changes of Surfactants in Magnetized Water. *J. Korean Chem. Soc.* **53**(2), 125-132.

Cai R., Yang H., He J. & Zhu W., 2009. The effects of magnetic fields on water molecular hydrogen bonds. *J. Mol. Struct.* **938**(1–3), 15–19.

Amiri M. & Dadkhah A.A., 2006. On reduction in the surface tension of water due to magnetic treatment. *Colloids Surf. Physicochem. Eng. Asp.* **278**(1), 252-255.

Hołysz L., Chibowski M. & Chibowski E., 2002. Time-dependent changes of zeta potential and other parameters of in situ calcium carbonate due to magnetic field treatment. *Colloids Surf. Physicochem. Eng. Asp.* **208**(1), 231-240.

Chibowski E., Hołysz L. & Szcześ A., 2003. Adhesion of in situ precipitated calcium carbonate in the presence and absence of magnetic field in quiescent conditions on different solid surfaces. *Water Res.* **37**(19), 4685-4692.

Ranade S., 1990. Paramagnetic metal contents and water proton spin-lattice relaxation times in tissues. *Physiol. Chem. Phys. Med. NMR* **22**(1), 15-18.

Ponomarev O. & Fesenko E., 1999. [The properties of liquid water in electric and magnetic fields]. *Biofizika* **45**(3), 389-398.

Akopian S. & Aĭrapetian S., 2004. [A study of specific electrical conductivity of water by the action of constant magnetic field, electromagnetic field, and low-frequency mechanical vibrations]. *Biofizika* **50**(2), 265-270.

Kronenberg K., 2005. Magneto hydrodynamics: The effect of magnets on fluids GMX international.

Holysz L., Szczes A. & Chibowski E., 2007. Effects of a static magnetic field on water and electrolyte solutions. *J. Colloid Interface Sci.* **316**(2), 996-1002.

Bruns S., Klassen V. & Konshina A., 1966. Change in extinction of light by water after treatment in a magnetic field. *COLLOID J.-USSR* **28**(1), 129.

Parsons S., Judd S., Stephenson T., Udol S. & Wang B., 1997. Magnetically augmented water treatment. *Process Saf. Environ. Prot.* **75**(2), 98–104.

Higashitani K., Kage A., Katamura S., Imai K. & Hatade S., 1993. Effects of a magnetic field on the formation of CaCO 3 particles. *J. Colloid Interface Sci.* **156**(1), 90-95.

Fu W. & Wang Z., 1994. The new technology of concrete engineering. *Beijing Publ. House Chin. Archit. Ind.* 56–59.

Wang G. & Wu Z., 1997. Magnetochemistry and Magnetomedicine. *Publ. House Ordnance Ind. Beijing* 82-90.

Kney A. & Parsons S., 2006. A spectrophotometer-based study of magnetic water treatment: Assessment of ionic vs. surface mechanisms. *Water Res.* **40**(3), 517-524.

Coey J. & Cass S., 2000. Magnetic water treatment. *J. Magn. Magn. Mater.* **209**(1), 71–74.
Cullity B.D. & Graham C.D., 2011. *Introduction to magnetic materials*, John Wiley & Sons.

Liu B., Gao B., Xu X., Hong W., Yue Q., Wang Y. & Su Y., 2011. The combined use of magnetic field and iron-based complex in advanced treatment of pulp and paper wastewater. *Chem. Eng. J.* **178**, 232–238.

Yadollahpour A., Rashidi S. & Ghotbeddin Z., 2014. Electromagnetic fields for the treatments of wastewater: a review of applications and future opportunities. *J Pure Appl Microbio* **8**(5), 3711-9.

Johan S., Fadil O. & Zularisham A., 2004. Effect of magnetic fields on suspended particles in sewage. *Malay J Sci* **23**, 141-148.

Strašák L., Vetterl V. & Šmarda J., 2002. Effects of low-frequency magnetic fields on bacteria Escherichia coli. *Bioelectrochemistry* **55**(1), 161-164.

Starmer J.E., 1996. School of water sciences.

Lipus L.C., Ačko B. & Neral B., 2013. Influence of magnetic water treatment on fabrics' characteristics. *J. Clean. Prod.* **52**, 374–379.

Selim M., 2008. Application of magnetic technologies in correcting under ground brackish water for irrigation in the arid and semi-arid ecosystem. Presented at the The 3rd International Conference on Water Resources and Arid Environments

Hilal M.H. & Hilal M.M., 2000. Application of magnetic technologies in desert agriculture. II- Effect of magnetic treatments of irrigation water on salt distribution in olive and citrus fields and induced changes of ionic balance in soil and plant. *Egypt. J. Soil Sci.* **40**(3), 423-435.

Mostafazadeh-Fard B., Khoshravesh M., Mousavi S.-F. & Kiani A.-R., 2011. Effects of Magnetized Water and Irrigation Water Salinity on Soil Moisture Distribution in Trickle Irrigation. *J. Irrig. Drain. Eng.* **137**(6), 398–402.

Mohamed A.I. & Ebead B.M., 2013. Effect of magnetic treated irrigation water on salt removal from a sandy soil and on the availability of certain nutrients. *Int. J. Eng.* **2**(2), 2305-8269.

Stuyven B., Vanbutsele G., Nuyens J., Vermant J. & Martens J.A., 2009. Natural suspended particle fragmentation in magnetic scale prevention device. *Chem. Eng. Sci.* **64**(8), 1904-1906.

Cefalas AC, Sarantopoulou E, Kollia Z, Riziotis C, Dražic G, Kobe S, and Meden A (2010). Magnetic field trapping in coherent antisymmetric states of liquid water molecular rotors. Journal of Computational and Theoretical Nanoscience, 7(9): 1800-1805. https://doi.org/10.1166/jctn.2010.1544

Tijing LD, Kim HY, Lee DH, Kim CS, and Cho YI (2010). Physical water treatment using RF electric fields for the mitigation of CaCO3 fouling in cooling water. International Journal of Heat and Mass Transfer, 53(7-8): 1426-1437. https://doi.org/10.1016/j.ijheatmasstransfer.2009.12.009

Helmuth R. 1987. Fly ash in cement and concrete. Portland Cement Association, Illinois.

Fan M., Brown R., Wheelock T.D., Cooper A.T., Nomura M., Zhuang Y. 2005. Production of a complex coagulant from fly ash. Chem. Eng. J., 106, 269–277

Moufti A. 2003. Doctoral thesis. Faculty of Sciences, Chourai Doukkali University, El Jadida, Morocco.

Adamiec P., Benezet J.C, Benhassaine A. 2005. Relationship between a silico-aluminous fly ash and its coal. Ecole des Mines d'Alès 6, Avenue de Clavières 30319 ALES Cedex

Zalaghi A., Lamchouri B., Toufik H., Merzoukia M. 2018. Valorization of natural porous materials in the treatment of leachate from the landfill uncontrolled city of Taza. Journal of Materials and Environmental Science, 5(5), 1643-1652.

Wiesemuller, W.,

Graich A., Bellarbi A., Khaidar M., Monkade M., Laamyem A., Rhanim H., Zradba A.2020. Stabilization/solidification by Portland cement of sludge as a residue of filtered waste water from a cardboard plant. Journal of Hazardous, Toxic, and Radioactive Waste, DOI: 10.1061/(ASCE)HZ.2153- 445 5515.0000506

Parsons., 1998.J.W .Isolation of humic substances from soils and sediments 1965. Investigations into the fractionation of soil organic matter. Albrecht-Thiier Archive., 9: 419--436

Flaig ,W. , Isotopes and Radiation in Soil Plant Nutrition Studies, 1965 . 3-19.21.

Dabin ,B., 1976 .Method extraction and fractionation of humic matter from the soil .Application to some educational and agronomic studies in tropical soils . Cah .ORSTOM, ser, Pedol .VOL.XIV, 4: 287–297

Kononova ,M ., BELCHIKOVA ,.N , 1961.A stady of soil humus substances by fractionation –Pochvoved II (I.Soil fertility, 190).

Walkley ,A., BLACK ,A., 1934. An examination of the Degtjareff method for determining soil C organic matter , and a proposed modification of the chromic acid–titrationmethode–soilsci .,PP.29–38

Anne, P., 1966. on the rapid determination of organic carbon in soils, Ann.

Li, L., Fan, M., Brown, R.C., Hans Van, L.J., Wang, J., Wang, W., Song, J. and Zhang, P. (2006) Synthesis, properties, and environmen-tal applications of nanoscale iron-based materi-als: a review. Crit. Rev. Environ. Sci. Technol. 36(5), 405– 431

Murad. S. (2006) The role of magnetic fields on the membrane-based separation of aqueous electrolyte solutions. Chem. Phys. Lett. 417, 465-470.

Guo, B., Han, H.B., and Chai, F. (2011) Influence of magnetic field on microstructural and dynamic properties of sodium, magnesium and calcium ions. Trans. NonferrousMet. Soc. China, 21, 494-498.

Wang, S.S.S., Chang, M.C., Chang, H.C., Chang, M.H., and Tai, C.Y. (2012) Growth be-haviour of aragonite under the influence of magnetic field, temperature, and impurity. Ind. Eng. Chem. Res. 511, 041-1049.

Seyfi A, Afzalzadeh R, and Hajnorouzi A (2017). Increase in water evaporation rate with increase in static magnetic field perpendicular to water-air interface. Chemical Engineering and Processing-Process Intensification, 120: 195-200. https://doi.org/10.1016/j.cep.2017.06.009

Chibowski E, Hołysz L, and Rafalski P (2011). Effects of static magnetic field on water at kinetic condition. Chemical Engineering and Processing: Process Intensification, 50(1): 124-127. https://doi.org/10.1016/j.cep.2010.12.005

Nakagawa J, Hirota N, Kitazawa K, and Shoda M (1999). Magnetic field enhancement of water vaporization. Journal of Applied Physics, 86(5): 2923-2925. https://doi.org/10.1063/1.371144

Amor HB, Elaoud A, Salah NB, and Elmoueddeb K (2017). Effect of magnetic treatment on surface tension and water evaporation. International Journal of Advance Industrial Engineering, 5: 119-124. https://doi.org/10.14741/Ijae/5.3.4

Mghaiouini R, Elaoud A, Garmim T, Belghiti ME, Valette E, Faure CH, and El Bouari A (2020a). The electromagnetic memory of water at kinetic condition. International Journal of Current Engineering and Technology, 10: 11-18.

Ueno S (2012). Studies on magnetism and bioelectromagnetics for 45 years: from magnetic analog memory to human brain stimulation and imaging. Bioelectromagnetics, 33(1): 3-22. https://doi.org/10.1002/bem.20714**PMid:22012916**

Dissado LA and Hill RM (1989). The fractal nature of the cluster model dielectric response functions. Journal of Applied Physics, 66(6): 2511-2524. https://doi.org/10.1063/1.344264

Kaden H, Königer F, Strømme M, Niklasson GA, and Emmerich K (2013). Low-frequency dielectric properties of three bentonites at different adsorbed water states. Journal of Colloid and Interface Science, 411: 16-26. https://doi.org/10.1016/j.jcis.2013.08.025**PMid:24112835**

Chahid EG, El Moznine R, Zradba A, Dani A, El Melouky A, Idrissi L, and Belboukhari A (2013). Analysis of relaxation processes and low frequency dispersion in waste water. Journal of Optoelectronics and Advanced Materials, 15(11-12): 1209-1216.

Laurati M, Sotta P, Long DR, Fillot LA, Arbe A, Alegrıa A, and Colmenero J (2012). Dynamics of water absorbed in polyamides. Macromolecules, 45(3): 1676-1687. https://doi.org/10.1021/ma202368x

Bonora PL, Deflorian F, and Fedrizzi L (1996). Electrochemical impedance spectroscopy as a tool for investigating underpaint corrosion. Electrochimica Acta, 41(7-8): 1073-1082. https://doi.org/10.1016/0013-4686(95)00440-8

Reuvers NJW, Huinink HP, Fischer HR, and Adan OCG (2012). Quantitative water uptake study in thin nylon-6 films with NMR imaging. Macromolecules, 45(4): 1937-1945. https://doi.org/10.1021/ma202719x

Weisenberger LA and Koenig JL (1989). NMR imaging of solvent diffusion in polymers. Applied Spectroscopy, 43(7): 1117-1126. https://doi.org/10.1366/0003702894203453

A. Elmelouky, A. Mortadi, R. El Moznin, E. Chahid, R. Lahkale, Dielectric spectroscopy studies in Zn2FeNO3, Zn2AlNO3 and Zn2Fe0.5Al0.5NO3, Int. J. Adv. Res., **2015**, 3,
465-475.

Szkatula, M. Balanda, M. Kope, Magnetic treatment of industrial water. Silica activation, Eur. Phys. J. Appl. Phys., **2002**, 18, 41-49.

M. Zielinski, A. Cydzik-Kwiatkowska, M. Zielinska, M. D. bowski, P. Rusanowska, J. Kopanska, Nitrification in activated sludge exposed to static magnetic field. Water Air Soil Pollut., **2017**, 228, 126.

C. Quinn, C. T. Molden, C. H. Sanderson, Magnetic treatment of water prevents mineral build-up, Iron and Steel Engineer, **1997**.

S.T. Bramwell, Ferroelectric ice, Nature 397 (1999) 212-213.

H. Xueyun, P. Yufeng, M. Zhongjun, Effect of magnetic field on optical features of water and KCl solutions, Optik (2016), https://doi.org/10.1016/j. ijleo.2016.04.096.

C.S. Choe, J. Lademann, M.E. Darvin, Depth profiles of hydrogen bound water molecule types and their relation to lipid and protein interaction in the human stratum corneum in vivo, Analyst 141 (2016) 6329-6337.

H. Xueyun, P. Yufeng, M. Zhongjun, Effect of magnetic field on optical features of water and KCl solutions, Optik (2016), https://doi.org/10.1016/j. ijleo.2016.04.096.

Cefalas, A.C., Sarantopoulou, E., Kollia, Z., Riziotis, C., Dražic, G., Kobe, S., Stražišar,J. , Meden, A.(2010) Magnetic field trapping in coherent antisymmetric states of liquid water Molecular rotors. J. Comput. Theor.Nanosci. 7, 1800-1805. DOI: 10.1166/jctn.2010.1544.

Florez M., Carbonell M.V. et Martinez E. (2007). Exposure of Maize Seeds to Stationary Magnetic Fields: Effects on Germination and Early Growth. Environ. Exp. Bot. 59:68-75.https://doi.org/10.1016/j.envexpbot.2005.10.006

O. Cheikh, Anis Elaoud, H. Ben Amor and M. Hozayn (2018), "Effect of permanent magnetic
Field on the properties of static water and germination of cucumber seeds". International

Journal of Multidisciplinary and Current Research. Pp108-116 V6. doi.org/10.14741/ijmcr.v6i01.10916

Flórez M., Martínez E., Carbonell M V. (2012). Effect of Magnetic Field Treatment on Germination of Medicinal Plants Salvia officinalis L. and Calendula officinalis L. Polish Journal Environmental Studies, 57-63.

Shabrangi A. et Majd A. (2009). Effect of Magnetic Fields on Growth and Antioxidant Systems in Agricultural Plants. PIERS Proceedings, Beijing, (China), March, 23-27.

J. Podleœny, S. Pietruszewski, and A. Podleœna. Efficiency of the magnetic treatment of broad bean seedscultivated under experimental plot conditions. International Agrophysics . 2004, 18, 65-71

María VictoriaCarbonellPadrino,ElviraMartinez,JasminAmaya. Stimulation of germination in rice (Oryza Sativa L.) by a static magnetic field. Electromagnetic Biology and Medicine 19(1):121-128. July 2000.DOI: 10.1081/JBC-100100303

Florez M., Carbonell M.V. et Martinez E. (2007). Exposure of Maize Seeds to Stationary Magnetic Fields: Effects on Germination and Early Growth. Environ. Exp. Bot. 59:68-75.https://doi.org/10.1016/j.envexpbot.2005.10.006

Moon, J.D., Chung, H.S. (2000) Acceleration of germination of tomato seed by applying ac elec-tric and magnetic fields. J. Electrost. 48(2), 103- 14.

Kataria, S., Baghel, B., Guruprasad K.N. (2017) Pretreatment of seeds with static magnetic field improves germination and early growth charac-teristics under salt stress in maize and soybean. Biocatalysis and Agricultural Biotechnology. 10, 83-90.

Abobatta, W. (2015b) Growth and fruiting of Valencia orange trees. Lambert Academic Pub-lishing (LAP). 196p.

Hachicha, M., Kahlaoui, B., Khamassi, N., Misle,E. ,Jouzdan, O. (2016) Effect of electro-magnetic treatment of saline water on soil and crops. Journal of the Saudi Society of Agricul-tural Sciences. 17(2), 154-162.

Silva,I.B. , QueirozNeto, J.C., Petri, D.F.S. (2014) The effect of magnetic field on ion hy-dration and sulfate scale formation. Colloids Surf. Physicochem. Eng. Asp. 465(1), 175-183.

Raiteri, P., Gale, J.D. (2010)Water is the key to nonclassical nucleation of amorphous calcium carbonate. J. Am. Chem. Soc. 132(49), 17623- 17634.

Colic, M.A., Chien, A., Morse, D. (1998) Syn-ergistic application of chemical and electromag-netic water treatment in corrosion and scale pre-vention. Croatica Chemica Acta. 71, 905-916.

Mghaiouini, R., Elmlouky, A., Salah, M., Al- Antary, T., Monkade, M., El Bouari, A.,Ghidan, A. (2020) Effect of electromagnetic fields on the pH of water under kinetic conditions. Fresen. Environ. Bull. 29, 7922-7933.

Shiyab, S., Al-Antary, T., Kafawin, O., Sowan, J., Akash, M. (2020) The effect of magnetic wa-ter on some characteristics growth in cucumber (Cucumis sativusL.). Fresen. Environ. Bull. 29, 6283-6291.

Kronenberg, K.J. (2005) Magnetohydronamies: The effect of magnets on fluids. GMX interna-tional.

Barefoot, R.R. and Reich, C.S. (1992) The cal-cium factor. The scientific secret of health and youth. Southeastern.PA. Triad marketing.5th edition. 90 -95.

Tkachenko, U.P.. (1997) Hydro magnetic aero ionizer in the system of spray, method of irriga-tion of agricultural crops. Hydromagnetic sys-tems and their role in creating micro-climate. In: Tkachenko,U.P.ed.) Practical Magnetic Tech-nologies in Agriculture. Dubai.

Nasher, S.H. (2008) The effect of magnetic wa-ter on growth of chick-pea seeds. Eng. and Tech. 26(9), 4p.

Celik, O., Atak, C. and Rzakulieva, A. (2008) Stimulation of rapid regeneration by a magnetic field in paulownia node cultures. Journal of Central European Agriculture. 9(2), 297-303.

Qodos, A.M.S.A., and Hozayn, M. (2010) Magntic water technology, a novel tool to in-crease growth, yield, and chemical constituents of lentil (Lens esculenta) under greenhouse con-dition. American Eurasian Journal of Agricul-tural and Environmental Science, 7(4), 457-462.

Mghaiouini, R. (2020a) Elaboration and physico-mechanical characterization of a new eco-mor-tar composite based on magnetized water and flay ash.Journal of Ecological Engineering. 21(4), 245-254.

Mghaiouini, R., Elmlouky, A., El Moznine, R., Monkade, M., El Bouari, A. (2020b) The Influ-ence of the electromagnetic field on the electric properties of water. Mediterranean Journal of Chemistry. 10(5), 507-515.

Mghaiouini, R., Benzbiri, N., Belghiti, M.E.,Belghiti, H.E. Monkade, M.,El Bouari, A. (2020c) Optical properties of water under the action of the electromagnetic field in the infra-red spectrum. Materials Today: Proceedings. https://doi.org/10.1016/j.matpr.2020.04.518.

Aarfane A. , Salhi A., El Krati M., Tahiri S., Monkade M., Lhadi E.K., Bensitel M. 2014. Kinetic and thermodynamic study of the adsorption of Red195 and Methylene blue dyes on fly ash and bottom ash in aqueous medium. J. Mater. Environ. Sci., 5(6), 1927-1939.

Su N., Wu Y., Mar C. 2000. Effect of magnetic water on the engineering properies of concrete containing granulated blast-furnace slage. Cement and Concrete Research, 30(4), 599-605.

Madsen H.E.L. 2004. Crystallization of calcium carbonate in magnetic field in ordinary and heavy water. Journal of Crystal Growth, 267(1-2), 251-255

Saddam M. and Ahmed M. 2014. Effect of magnetic water on engineering properties of concrete. Al- Rafedain Engineering, 17(1), 77-82.

Joshi K.M. and Kamet P.V.1966. Effect of magnetic field on the physical properties of water. J.Ind. Chem.Soc. 43, 620-622.

Abdel-Magid TIM, Hamdan RM, Abdelgader AAB, Omer MEA, Ahmed NMR. Effect of magnetized water on workability and compressive strength of concrete. Procedia Engineering. 2017, 193, 494–500. DOI: 10.1016/j.proeng.2017.06.242.
An-Tai Ma. Effect of magnetic water on engineering properties of self - compacting concrete with waste catalyst. A thesis submitted to institute of construction National Yunlin University of Science and Technology in partial fulfillment of the requirement for degree of Master of Design in Construction

Engineering Taiwan-Republic of China. 2007. DOI: 10.13140/ RG.2.2.35245.38880.

Fargas F, TouZePh. The IRTF spectrometer, an interesting method for the characterization of cements. Bulletin from the Bridges and Carriageways Laboratories. 2001, 77–88.

P. Asokan, S. Mohini et S. R. Asolekar, "Coal combustion residues - environmental implications and recycling potentials", Resources, Conservation and Recycling, pp. 43, 239-262, 2005. No. 9, pp. 30 - 32, September, 2004.

K. Haldun et K. Mine, "Usage of coal combustion bottom ash in concrete mixture",
Construction and Building Materials, pp. 1922-1928, 2007.

ANNEXES

Scientific work during the thesis

Poster communications :

R.mghaiouini, Contribution to the study of the thermomechanical properties of mortars based on biyada lime and seaweed,2019 , the first international day of energetics and materials,faculte des science eljadida

1. R.Mghaiouini,A.Elmelouki,A.Graich,T.Garmim,M.Monkade,A.Elbouari,J .Elhajri,A.Zradba,K.Kandoussi.A.loaurdi,2019,determination of humic and fulvic acids from manufactured compost based on manure and marketd coffee, optimization of extraction methods to adapt them to the manufacture of new organic fertilizers,2019,7 eme symposium international des plantes aromatius et medicinales...

Oral presentations:

- R.Mghaiouini, T.Garmim,T M.Monkade ,A.Elbouari; influence of electromagnetic field on water pH at kinetic conditions.2019,internatiobnal congress on advanced materials for photonics sensing and energy conversion energy application AMPSECA 2019.marrakech

3- R.Mghaiouini, T.Garmim,M.Monkade ,A.Elbouari,The electromagnetic memory of water at kinetic condition, th 1 ere edition of the international conference on renvelable energy and applications 2019,ENIS casablanca

4- R.Mghaiouini, T.Garmim,M.Monkade ,A.Elbouari,2019, Physicochemical property monitoring of magnetized water by impedance and dielectric spectroscopy,Systeme et procedes industriels ENSA eljadida

R.Mghaiouini,M.Salah,M.Monkade,A.El bouari, A New knowledge of water magnetism phenomenon, n the 1st international E-conference on Climate Nexus Perspectives (I2CNP): Water Food and Biodiversity organized by the research team "Environment and Natural Resources Management" in the Higher School of Technology,Khenifra,inJune,04th,2020.

INDEXED ARTICLES :

1-Mghaiouini.,R(2020).Elaboration and Physico-Mechanical Characterization of an

New Eco-Mortar Composite Based on Magnetized Water and Flay Ash". Journal of

Ecological Engineering 21 no. 4: 245- 254.doi:10.12911/22998993/119816. (scopus)

2-Mghaiouini R, Elmelouky A, Monkade M and Elaoud A et al. (2020). Physicochemical property monitoring of magnetized water by impedance and dielectric spectroscopy in the kinetic condition. International Journal of Advanced and Applied Sciences, 7(8): 91-104. https://doi.org/10.21833/ijaas.2020.08.010 (Web of science)

3- Mghaiouini.R., Elmlouky.A.,El Moznine.R.,Monkade.M.,El Bouari.A. (2020) The Influence Of The Electromagnetic Field On The Electric Properties Of Water. Mediterranean Journal of Chemistry, Vol 10 , No 5 (2020) . DOI: http://dx.doi.org/10.13171/mjc10502005181406rm (scopus et Pub Med)

4- Mghaiouinia R., Benzbiria.N., Belghiti.M.E., Belghitie.M.E., Monkade.M., El bouari .A . (2020) Optical properties of water under the action of the electromagnetic field in the infrared spectrum. Materials Today: Proceedings. https://doi.org/10.1016/j.matpr.2020.04.518 (scopus)

5- Redouane Mghaiouini, Anis Elaoud, Toufik Garmim, M. E.Belghiti , Eric Valette,Charles Henri Faure, Mahmoud

Hozayn,MohamadeMonkade and Abdeslam El Bouari,The Electromagnetic Memory of Water at Kinetic Condition.

International Journal of Current Engineering and Technology ,DOI: 10.14741/ijcet/v.10.1.3lin, i.

6- R. Mghaiouini, A. Elaouad, H Taimoury, I. Sabir, F. Chibi, M. Hozayn, T Garmim, R Nmila, H Rchid, M . Monkade, A

El Bouari,Influence of the Electromagnetic Device Aqua 4D on Water Quality and Germination of Lettuce (Lactuca sativa L.), International Journal of Current Engineering and Technology, 2020,Vol.10, No.1. DOI: https://doi.org/10.14741/ijcet/v.10.1.4

7- Redouane Mghaiouini, Abderrazzak Graich, Anis Elaoud, Toufik Garmim, M.E. Belghiti, Nisrine Benzbiria, Mahmoud Hozayn, Mohamade Monkade and Abdeslam El Bouari,Formulation and Physico-mechanical Characterization of an Eco-mortar Composite Based on Bottom Ash and Magnetized Water. Indian Journal Of Science and Thechnology, March 2020, Vol 13(10), 1172 - 1187 DOI: 10.17485/ijst/2020/v13i10/149889 (W

8- M.E. Belghiti, M. Mihit, A. Mahsoune, A. Elmelouky, R. Mghaiouini,A. Barhoumi, A. Dafali, M. Bakasse, M.A. El Mhammedi, M. Abdennouri, Studies of Inhibition effect "E & Z" Configurations of hydrazine Derivatives on Mild Steel Surface in phosphoiric acid, Journal of Materials Research and Technology ; Correhttps://doi.org/10.1016/j.jmrt.2019.09.051 2238-7854

9-Redouane Mghaiouini, Abderrahmane Elmlouky2, Mohammed Salah3, Tawfiq M Al-Antary4, *, Mohamed Monkade2, Abdeslam El Bouari1, Alaa Y Ghidan.EFFECT OF ELECTROMAGNETIC FIELDS ON THE ph OF WATER UNDER KINETIC CONDITIONS.July 2020.Fresenius Environmental Bulletin 29(9):7922-7933

10- Redouane Mghaiouini, Tawfiq M Al-Antary, Safwan Shiyab, Ibtissam Sabir, Mohammed Salah, Rachid Nmila, Halima Rchid, Mohamed Monkade, Abdeslam El Bouari. POTENTIAL IMPACT OF MAGNITIZED WATER ON PLANT GROWTH OF LETTUCE, LACTUCA SATIVA L. Volume 29 - No. 09/2020 pages 7891-7900 Fresenius Environmental Bulletin 7891

11- Redouane Mghaiouini,Mohammed Salah, Mohamed Monkad and Abdeslam El Bouari .
Effect of physicochemical parameters on magnetic treatment of water. August 2020 E3S Web of Conferences 183:05001. DOI: 10.1051/e3sconf/202018305001

12- INVESTIGATE THE EFFECTS OF SALINITY AND TEMPERATURE ON A ELECTROMAGNETIC WATER in review. Journal COLLOIDS AND SURFACES COLSUB-D-20-01672 .

13-Formulation of new biostimulant of plant and soil correction, correction terminer. Journal HELIYON-D-20-01967.

14- Effect of the Low Frequency Electromagnetic Field on the Mechanical Properties of the Mortar Elaborated by Fly Ash, Bottom Ash and Magnetized Water.in review journal KEY ENGINEERING MATERIALS

BREVET

International patent application to the EUROPEAN PATENT OFFICE (EPO) under the title formulation and characterisation of a new ecomortier based on magnetised water and fire ash.

Printed by Books on Demand GmbH, Norderstedt / Germany